Symposia
Biologica
Hungarica
27

Symposia Biologica Hungarica

Vol. 27

Redigunt

H. KALÁSZ et L. S. ETTRE

AKADÉMIAI KIADÓ, BUDAPEST 1985

CHROMATOGRAPHY
THE STATE OF THE ART

Proceedings of the Budapest Chromatography Conference
June 1–3, 1983
Budapest, Hungary

Edited by
H. KALÁSZ
and
L. S. ETTRE

Volume II

AKADÉMIAI KIADÓ, BUDAPEST 1985

53738159

chem

ISBN 963 05 4081 9 (Vols I—II)
ISBN 963 05 4089 4 (Vol. II)

Printed in Hungary

CONTENTS

VOLUME I

Preface   v

KNOLL, J.
        Introductory remarks   1

THEORETICAL ASPECTS

HORVÁTH, CS.
        Quo Vadis HPLC   11

PUNGOR, E., HORVAI, G., FEHÉR, ZS. and TÓTH, K.
        Electrochemical detection in chromatography   19

BEREZKIN, V.G., KOLOMIETS, L.N. and KOROLEV, A.A.
        Development of transport detectors for liquid
        chromatography   35

GLÖCKNER, G.
        HP precipitation chromatography - A new technique
        for the elucidation of the molecular structure of
        copolymers   41

EKSTEEN, R. and THOMA, J.J.
        Evaluation of short and efficient reversed-phase
        columns for the fast analysis of theophylline by
        HPLC   51

WULFSON, A.N. and YAKIMOV, S.A.
        HPLC of nucleotides. II. General methods and their
        development for analysis and preparative separa-
        tion. The approach to selectivity control   63

OBRUBA, K.
        Analysis of products of anthracene sulphonation
        by reversed-phase high-performance liquid
        chromatography   107

GÖNDÖS, GY. and ORR, J.C.
    Separation of steroid ketoalcohols and diols by
    high-performance liquid chromatography   113

OHMACHT, R. and TÓTH, GY.
    Determination of carotenoids and retinol in
    human serum   119

STATIONARY PHASES FOR CHROMATOGRAPHY

OHMACHT, R. and MATUS, Z.
    New "Chromsil" packings   129

KICINSKI, H.G., MAASFELD, W. and KETTRUP, A.
    Synthesis, characterization and analytical applica-
    tion of chemically modified silica in HPLC   143

TOGUZOV, R.T. and TIKHONOV, Yu. V.
    Natural chromatographic systems in biological
    objects   153

CHROMATOGRAPHY OF AMINES AND AMINO ACIDS

SEILER, N., KNÖDGEN, B. and BARTHOLEYNS, J.
    Polyamines and their derivatives as markers of
    neoplastic growth   163

BARDÓCZ, S., KARSAI, T. and ELŐDI, P.
    Determination of polyamines by overpressured
    thin-layer chromatography (OPTLC)   187

KERECSEN, L., KALÁSZ, H., TARCALI J., FEKETE, J.
and KNOLL, J.
    Measurement of DA and DOPAC release from rat
    striatal preparations "in vitro" using HPLC with
    electrochemical detection   195

HÁRSING, L.G. Jr., TEKES, K., MAGYAR, K. and VIZI, E.S.
    In vitro effect of l-deprenyl on dopaminergic and
    cholinergic neural transmissions in caudate
    nucleus of the rat   203

HÁRSING, L.G. Jr., TÁRCZY, M., BIHARI, K. and VIZI, E.S.
    Measurement of plasma l-dopa level by high-
    performance liquid chromatography coupled with
    electrochemical detection in Parkinsonian patients
    treated with Madopar®   219

SEPARATION OF PEPTIDES AND PROTEINS

DAVILA-HUERTA, G. and VARGA, J.M.
    Identification of positional analogues of SPDP-β-
    MSH derivatives by reversed-phase high-performance
    liquid chromatography of dansyl amino acids   227

REICHELT, K.L. and EDMINSON, P.D.
Principles in isolation of peptide factors from
urines with retained biological activity. Example
from schizophrenia   237

SZÓKÁN, GY., PENKE, B., BALÁSPIRI, L. and TÖRÖK, A.
RP-HPLC in the analysis, synthesis and reactions
of neuropeptides and their derivatives   257

HUDECZ, F. and SZÓKÁN, GY.
Structure analysis of branched chain poly- and
isopeptides based on HPLC of their dansyl
derivatives   273

TÓTH, G., PENKE, B., JANÁKY, T., KOVÁCS, K. and
RIVIER, J.
Reverse phase HPLC of hydrophobic and hydrophilic
modified peptide hormones   287

UI, N.
Application of high-performance liquid chromatography for the rapid estimation of the molecular
weights of proteins   299

PASECHNIK, V.A., ANDREEV, S.V., ISHCHENKO, A.M.,
PIVOVAROV, A.M. and KOROBITSIN, L.P.
Chromatographic and functional analysis of
complement proteins in α-interferon preparations   311

SZABÓ, L. and MATKOVICS, B.
Novel application of superoxide dismutase   331

NAGY, J., MAZSAROFF, I., VÁRADY, L. and KNOLL, J.
Studies on the purification and properties of
satietin, an anorexogenic glycoprotein of
biological origin   337

BÁRDOS, L.
Isolation of the retinol binding protein (RBP)
from the serum of farm animals   359

GUOTH, J., SÁROSI, P., IDEI, M., MENYHÁRT, J. and
PAJOR, A.
Inhibition of myometrial contractions in vitro
by a substance in extract of sow ovaries. The
relation of the substance to standard relaxin   367

IDEI, M., GRÓF, J., GUOTH, J., PAJOR, A. and
MENYHÁRT, J.
Application of isolated organ preparations as
sensitive and selective detectors in the
practice of liquid chromatography   377

SEPARATION OF DRUGS AND METABOLITES

MAGYAR, K.
    Study of deprenyl metabolism   391

EKIERT, L., BOJARSKI, J. and MOKROSZ, J.
    Some aspects of structure-chromatographic
    behavior relationships in TLC of barbiturates   403

BIDLÓ-IGLÓY, M.
    Reversed-phase HPTLC and HPLC studies of some
    pyridazine derivatives   413

SUPRYNOWICZ, Z., LODKOWSKI, R., BUSZEWSKI, B. and
OCHYŃSKI, J.
    HPLC separation of bromfenvinphos and its
    metabolites in physiological samples   421

KURCZ, M. TÓTH, K. and PÁLOSI-SZÁNTHÓ, V.
    Bioequivalence study of Sensit® and Sensit-β-
    cyclodextrin complex by HPLC and radioisotopic
    methods   435

FODOR, I., FODOR, G., FODOR, M. and DINYA, Z.
    Characterization of Novicardin with chromato-
    graphic methods   445

URBÁN SZABÓ, K. and KURCZ, M.
    Application of micro-HPLC for the determination
    of therapeutic levels of drugs   459

SZEPESI, G. and GAZDAG, M.
    Optimization of chromatographic separations in
    the pharmaceutical analysis   467

VOLUME II

THIN-LAYER CHROMATOGRAPHY

STAHL, E.
    A quarter of a century of thin-layer chromato-
    graphy. An interim report   497

KALÁSZ, H.
    Forced-flow thin-layer chromatography   501

GANKINA, E.S., LITVINOVA, L.S., EFIMOVA, I.I.,
BASKOVSKY, V.E. and BELENKII, B.G.
    High-performance reusable plates for thin-layer
    chromatography and their application to the
    analysis of lipids and synthetic polymers   517

DÉVÉNYI, T. and KALÁSZ, H.
    Thin-layer ion-exchange chromatography in
    biochemical analysis   535

POLYÁK, B., ÁBRAHÁM, M. and BOROSS, L.
    Systematic amino acid and glutathione analysis
    by overpressured thin-layer chromatography   591

CSERHÁTI, T., BORDÁS, B., DÉVAI, M. and KUSZMANN, A.
    Application of principal component analysis to
    compare different thin-layer chromatographic
    systems   601

ROZYLO, J.K., MALINOWSKA, I. and PONIEWAZ, M.
    Some problems of thin-layer chromatography
    as a pilot technique for adsorption liquid
    column chromatography   615

VOLYNETS, M.P., RUBINSTEIN, R.N. and KITAEVA, L.P.
    Peculiarities in determining the parameters of
    the ion-exchange thin-layer-chromatographic
    system in the presence of complex formation   633

GAS CHROMATOGRAPHY

MANCAS, D.GH.
    Rapid gas-chromatographic determination of
    phenylacetic acid in the fermentation broth of
    benzyl penicillin production   655

MANCAS, D.GH.
    Analysis of thiodiglycolic acid in urine by
    capillary gas chromatography   659

MANCAS, D.GH.
    Gas-liquid chromatographic method for the
    determination of serum cholesterol by flash-
    heater silylation   665

BOROSS, F. and TÓTH-MÁRKUS, M.
    Isolation of and investigation into the ester
    fraction of fruit flavour concentrates   673

KALISZAN, R., LAMPARCZYK, H., PANKOWSKI, M.
DAMASIEWICZ, B., NASAL, A. and GRZIBOWSKY, J.
    A gas chromatographic polarity parameter and
    its application in studies of quantitative
    structure-olfactory activity relationships in   679
    phenols

BARCELÓ, D., GALCERÁN, M.T. and EEK, L.
    Comparison of the porous polymers chromosorb 101
    and Chromosorb 102 as supports coated with Ethofat
    60/25 in gas chromatography   697

CALCULATION AND OPTIMIZATION METHODS

BIAGI, G.L., BARBARO, A.M., GUERRA, M.C., CANTELLI
FORTI, G., BOREA, P.A., PIETROGRANDE, M.C.,
SALVADORI, S. and TOMATIS, R.
  Chromatographic parameters of lipophilicity of
  a series of dermorphin related oligopeptides   719

RAFEL, J.
  Comparison of optimization methods for a
  separation quality parameter in HPLC   731

VALKÓ, K.
  Microcomputer analysis of HPLC reversed-phase
  retention behavior of several pharmaceutical
  agents depending on the composition of the eluent   739

JÁNOS, É., BORDÁS, B., CSERHÁTI, T. and SIMON, G.
  Substituent effects in RPTLC   751

VARIOUS TOPICS

TOGUZOV, R.T., TIKHONOV, Yu. V., MEISNER, I.S.,
KOL'TSOV, P.A., POLYKOVSKAYA, O. Ya. and BUTOV, Yu. S.
  Chromatographic analysis of purine and pyrimidine
  derivatives from the tissues of animals and
  humans in experimental and clinical pathology   761

COPIKOVÁ, J., KVASNICKA, F. and MUSIL, V.
  Ion exchanger Ostion KS 4.2% and separation of
  saccharides   809

GAIKWAD, A.G. and KHOPKAR, S.M.
  Cation exchange separation of yttrium in mixed
  solvents

LEHMANN, H.
  Chromatographical determination of abscisic acid
  and its glucosyl ester   833

BELENKII, B.G. GANKINA, E.S., KURENBIN, O.I. and
MALTSEV, V.G.
  Capillary liquid chromatography. Problems and
  prospects   841

List of contributors   887

Subject index   901

X

THIN-LAYER CHROMATOGRAPHY

A QUARTER OF A CENTURY OF THIN-LAYER CHROMATOGRAPHY, AN INTERIM REPORT

EGON STAHL

Department of Pharmacognosy and Analytical Phytochemistry
University of Saarlamdes, Saarbrücken, FRG

1958 marks the year of the beginning of the worldwide use
of thin-layer chromatography /TLC/ which was made possible by
the use of very small particle size material in the layers. It
was then shown that not only lipophilic mixtures could be separated but mixtures of cardiac glycosides, alkaloids, sugars
and amino acids as well. Fast and simple separations encouraged
later tests to use absorbants, which also have such narrow-ran-
ge small particles, in column chromatography. This accounted
for the need of high pressures in order to secure sufficient
flow rates as the columns were tightly packed. Thus, high-pressure or high-performance liquid chromatography /HPLC/ was born.
Instrumentally seen, you may think of gas chromatography   as
father and thin-layer chromatography as mother of the new analytical child.

Nowdays, we tend to classify the field of chromatography
according to the transport phase used: liquid, gas, or "electro"
chromatography. As the sample preparation usually requires more
time than the actual separation procedure, we developed   thermal
separation and application techniques which are fast to handle,
the TAS or the TFG have added significant advantages to   TLC.
Working efficiently in the field of chromatography, it is most
advantageous to choose the "right" mobile" phase, stationary
phase, and separation technique.

As we advanced in the knowledge of chromatography, gradient
thin-layer chromatography came along and offered new possibilities. Gradient layers make it possible for instance to obtain

selective information concerning the individual components of a mixture. Also, important focussing effects - analogously to the phenomenon of isoelectric focussing - with narrowing of the bands could be achieved which lets more components being seen on one layer. Further advances were reached with centrifugal TLC and also sequential TLC.

In the first tests with TLC, standard conditions were set up. After so many years, one asks oneself whether standard conditions, formulated 25 years ago, can still be valid today. There have been variations with diminution and enlargements, yet we must truthfully say that the proven standard conditions are still the most widely used.

Every chromatographic method is only as efficient as its detection of the separated substances. There are destructive and non-destructive methods available in TLC. There are also numerous detection reactions which can be divided into three groups: universal detectors, group specific detectors and substance specific detectors, and one has to decide on the priority of the expectation, be it specificity, sensitivity, reproducibility or convenience.

Right from the beginning of the extensive use of TLC we aimed for reliable quantitative evaluation of the obtained chromatograms. Very early in the 1960, special photometers were manufactured for TLC and further developed. In the meantime, computer controlled evaluation has found its way in TLC also. It seems that with sample removal and quantitative evaluation being an integral part of the instrument in HPLC this will be a prominent method in the future. In discussing the future of TLC one should bear in mind the following points:

1. TLC offers the greatest freedom in the easy choice of stationary and mobile phases.

2. It offers a large number of different development techniques.

3. It offers the largest number of methods of detection.

4. It is possible to separate and visualize many samples and reference mixtures simultaneously.

5. Simple methods are available for combining and coupling
    with microextraction and identification techniques.

6. Thin-layer chromatography is the simplest and most econ-
    omical chromatographic technique for fast separation
    and visual evaluation.

These advantages justify the statement that TLC is here
to stay being an extraordinarily useful method. Also, it is a
method that can be coupled and combined with other methods in
order to secure the best separation. In coupling, all standard
TLC methods can be used.

The unabbreviated lecture was puslished in "Angewandte
Chemie, International Edition in English", Volume 22, Number 7,
July 1983, pages 507-516.

FORCED-FLOW THIN-LAYER CHROMATOGRAPHY

HUBA KALÁSZ

Department of Pharmacology, Semmelweis University of Medicine,
H-1445 Budapest, P.O.B. 370, Hungary

Liquid chromatographic separations are realized in two different geometrical formations of the stationary phase bed, either column or plane. These arrangements have special characteristics, so the results of the separations are also different even when the stationary phase, the mobile phase and the sample to be separated are the same.

In liquid column chromatography /LCC/, the stationary phase is packed in a cylindrical tube and the sorbent should be extensively washed and equilibrated with the mobile phase before the first sample is separated. As the bed in LCC is not in contact with the vapor of the mobile phase, the separation process is dominated by the triple equilibrium among the solute-solvent-sorbent as it is demonstrated in Fig. 1. Both in classical column chromatography and in its high-performance variation, the solvent flow velocity is controlled by pumps which supply the mobile phase with a constant flow rate during the whole period of the chromatographic separation. This is one of the reasons why the elution volume of each component is definitely proportional to the elution time in separate experiments and the results can be repeated without significant deviations. In LCC, the effluent is generally monitored by UV detectors; the use of flow cells and precise instrumentation makes reliable quantification possible. As the consequence of the nature of LCC, the samples can be separated one by one. The stationary phase can be used several times for separation but parallel experiments can be performed successively only in space or time.

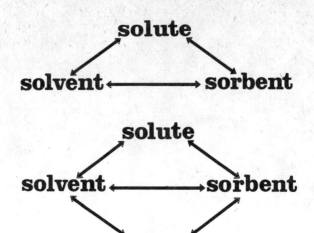

Fig. 1. Equilibria play the main role in column liquid
chromatography /top/ and thin-layer chromatography
/bottom/.

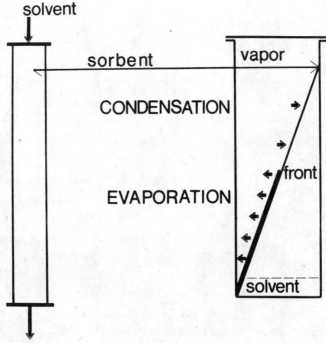

Fig. 2. Column liquid chromatography /left side/ is done using
a closed bed. Thin-layer chromatography /right side/
is realized with an open stationary phase where the
evaporation and condensation of the mobile phase are
superimposed on the actual chromatographic process.

At the same time, both analytical and preparative procedures are possible, but the realization of two-dimensional chromatography needs the distinct process for each fraction. In addition, there is not any well-defined method to detect the substance/s/ remaining at the very beginning of the column in consequence of its very moderate mobility.

In planar technique, the stationary phase is spreaded on a plate /glass, plastic, etc./ and the samples are spotted on the dry sorbent. The thin-layer plate is placed into a chamber which contains the mobile phase at its bottom and the space is more or less saturated with the vapor of the solvent. The thin-layer of the sorbent is in contact with the vapor of the mobile phase, i.e., possibility is provided for the constant precipitation and evaporation of the solvent /Fig. 2/. Thus, there are vapor-solvent and vapor-sorbent equilibria which are superimposed on the actual chromatographic processes. In addition, the stationary phase is dry at the beginning of the chromatographic separation; therefore the mobile phase undergoes frontal chromatography. These two procedures /i.e., frontal chromatography and continuous evaporation and condensation of the solvent vapor/are responsible for the fact that in multicomponent solvent systems, fronts of different compositions can be observed. These fronts are called α, β, γ, etc. fronts, and their arrangement depends on the physical circumstances of thin-layer chromatography, the saturation of the chamber, the activity of the sorbent, etc. The mechanisms taking place at one solvent component seem to be much simpler although the basic picture of the thin-layer chromatogram remains the same. In dichloromethane a diffuse solvent front is moving forward on the silica gel plate and the position and separation of the multi-component dye-mixture is essentially influenced by the saturation-unsaturation circumstances of the chamber as it can be seen in Fig. 3. The constant evaporation of the solvent from the layer causes that the spots near to the mobile phase front /where the loss of the solvent is more expressed/ are rather crowded to each other, if unsaturated chambers are used. On the contrary, if saturated chamber is used, the separated spots can be observed far enough from each other. Naturally, the situa-

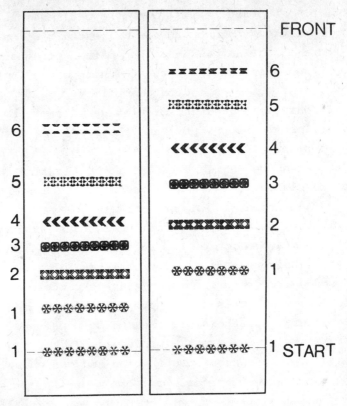

Fig. 3. The separation of some dye-substances gives different
pictures in saturated and unsaturated chambers.

tion shows an opposite trend closer to the start line: therefore the saturated chamber can be advantageous for the spots
near to the solvent front while the non-saturated chamber is
preferred in the separation of substances moving with low R_f
values.

The difficulties of reproducibility due to the constant
evaporation and condensation of the mobile phase can be almost
entirely eliminated by covering the sorbent layer with a glass
plate or membrane. This idea was originally suggested by Tyihák and Held /1/ who called the method thin-layer chromatography
with ultramicro chamber. The covering glass plate increased
the reproducibility of TLC but somehow the use of ultramicro
/UM/ chambers has never spread into everyday practice.

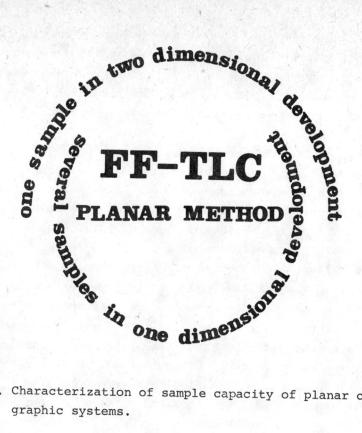

Fig. 4. Characterization of sample capacity of planar chromatographic systems.

On the other hand, thin-layer chromatography /TLC/   has
several advantages over the column methods /Fig. 4/. Several
samples can be investigated in a parallel way, the separated
spots can easily be detected by specific and sensitive color
reactions or the separation itself can be continuously followed
if dye-components or substances with intensive color are chromatographed. TLC does not need intensive and expensive instrumentation. As for the mobile phases, strong acids and bases
can also be used because both the all-glass chambers and the
silica gel on the glass plate have enough resistance.

TLC makes two-dimensional chromatography much easier and
its application has been really wide spread as detailed in a
recent study /2/. However, the above-mentioned shortcomings of
TLC have resulted in a constant drive to eliminate them pre-

serving at the same time its several advantages. This intent and effort has always been rather forced by the example, how column chromatography has been developed in the last 20 years.

The development of column liquid chromatography resulted in the so-called high-performance liquid chromatography /HPLC/. Some basic aspects of HPLC are: the use of spherical, homodisperse, microparticulate particles, application of a pump for continuous delivery of the mobile phase with constant flow rate, detection of the effluent by micro flow cells, precise sampling systems, the use of reversed-phase materials, and the application of specific detection devices /e.g. fluorimetric monitors, electrochemical detectors, etc./. By this way, HPLC can result in very high theoretical plate numbers, the separation needs only a short time and quantitative evaluation is precise and reliable even when measuring substances below 100 pg.

The inventors of the new methods of thin-layer chromatography have adopted step by step most of the new achievements of HPLC developments /Fig. 5/. Some ideas and arrangements of controlling and adjusting the flow rate were published by Brenner and Niederwieser /3/, Saunders and Snyder /4/, Hara and Mibe /5/, Soczewinski and Wawrzynowicz /6, 7/ and Determann /8/. The application of particles with small diameter resulted in a method /called HPTLC/ which gives good resolution within a short period of time and short developing distance. HPTLC requires the use of special plates, special sample application device and separation chambers as well as special detectors for quantitative evaluation. However, the real limit of HPTLC is not the instrumentation but the very limited distance for development. At longer distance of run, the movement of the solvent front becomes extremely slow.

Also Tyihák and Held /1/ did not render any gadget or well-manageable arrangement to cover the stationary phase, the simple overlay of the sorbent with a glass plate and linking them together with clips improves reproducibility. Fig. 6 shows this fact: the right-side chamber has non-saturated vapor phase while the left-side chamber has saturated vapor phase /by the help of a paper strip/, however, no remarkable differences can be observed between the behaviour of the dye-components separated in the chambers.

506

1. **covering the sorbent layer,**

2. **tight close of the sorbent layer,**

3. **forced flow of mobile phase,**

4. **arrangement of linear front.**

Fig. 5. The tracks of intentions which realized the forced-flow thin-layer chromatography with linear solvent front.

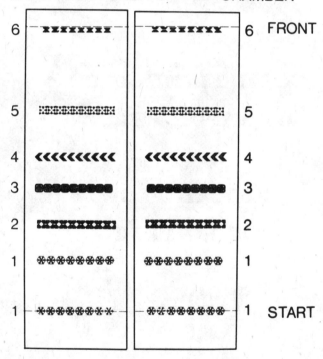

SATURATED- UNSATURATED-CHAMBER

Fig. 6. The chromatographic separations in saturated and unsaturated chambers give the same results if covered stationary phase is used.

The realization of the totally closed stationary phase bed in planar arrangement permits chromatography without vapor phase and with pump-delivered solvent supply. This system ensures the reproducibility of the R_f values, offers fast development with constant flow rate as well as the extension of the solvent front /Fig. 7/. This latter has never been a theoretical but a technical problem, as the overdevelopment of the solvent front can also be performed in conventional TLC chambers, however the necessary time is long enough, and the flow rate can be changed from time to time.

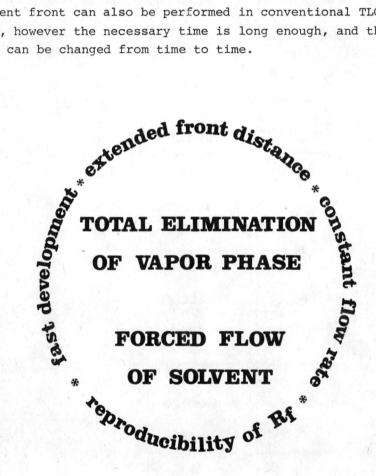

Fig. 7   Some features of forced-flow thin-layer chromatography with circular and linear solvent front.

Two forms of thin-layer chromatographs with totally closed sorbent layer have been developed /9-12/. One set-up forms circular solvent front and this apparatus /9/ gives the possibility of sampling during development. The other type of set-ups results in linear solvent front /9-12/, and sampling is possible before the start of development as it is usual in conventional TLC. Although circular TLC can also separate spots dried on the sorbent before development, the real feature of the circular technique is that various color reagents can be used if different segments of the chromatogram are treated. However, the basic form of the overpressured thin-layer chromatographs is the set-up which gives linear solvent front. The scheme of the arrangement is demonstrated in Fig. 8. The stationary phase bed is bordered with some material which cannot be penetrated by the mobile phase. At the beginning of our research we used candle wax; later the use of a polymerized plastic became usual. The solvent is delivered by a pump through a single-point solvent inlet. From this position, the mobile phase might go toward all directions but the bordering of the plate and the channel on the sorbent direct the solvent in a well-defined linear front shape toward the direction of development. The tight close of the covering membrane prevents flow over the sorbent layer, and the forced-flow solvent supply gives a constant and regulated flow rate.

From the technical point of view, the set-up and method can also be called overpressured thin-layer chromatography /OPTLC/ as the stationary phase bed has a planar arrangement /..TLC/ and it is tightly closed by a membrane with the help of an overpressure on it /OP.../. At the same time, the other aspect of the essence of the arrangement is that the solvent is not supplied by capillary forces but by a pump which is obviously "forced flow" and the pump can deliver a constant flow rate even when the resistance of the system is increased due to the longer solvent front, the increased viscosity and side reactions /total wetness of the layer/. From retrospective points of view of nomenclature, overpressured thin-layer chromatography seems to be rather useful /9-12/ but on the basis of

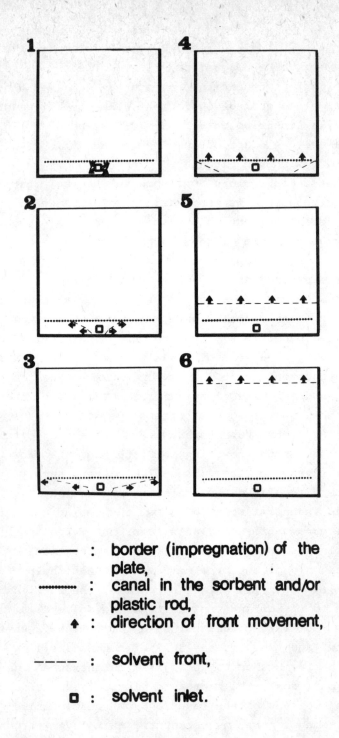

Fig. 8 The scheme of linear solvent front formation.

the recent experiments the future strongly suggests the use of Forced-Flow Thin-Layer Chromatography /FF-TLC/. The developing distance vs. developing time diagrams show definite decline in the case of TLC using conventional chambers but straight lines were obtained if the experiments with FF-TLC were carried out. This finding is very important especially from the point of view of reversed phase thin-layer chromatography.

The other argument favoring the use of the expression FF-TLC is that the covering of the plate /which is the essence of OPTLC/ does not mean the necessity of using any pump as the solvent movement can also be realized by the conventional method of touching the bottom of the layer into the solvent when the mobile phase will move also propelled by capillary forces; the "forced flow" expression means the elimination of the spreading of the mobile phase as well as that of the flow outside the sorbent layer by the application of the tight covering of the membrane.

Naturally, the use of our FF-TLC set-up is not restricted to the use of silica gel plates. In our earlier experiments /12/ Fixion 50X8 plates were also widely used and the experiments proved that the highly increased flow velocity of the solvent front improves the separation and optimal efficiency can be achieved at a relatively high flow rate. Similarly, alumina, cellulose and polyamide layers can also be used for FF-TLC just as the reversed-phase sorbents /13/.

Fig. 9 demonstrates the striking common characteristics, a declining form of the front distance vs. time diagram in the case of conventional chamber while FF-TLC gives straight lines.

Another feature of FF-TLC is that overdevelopment can easily be carried out. Polyák and coworkers detailed /14/ in the case of amino acids that the substances with low mobility can be advantageously investigated if the solvent front moved over the top of the TLC plate and the development is continued. In this case, the substance with R_f 0.1 - 0.3 can be placed in about the middle of the thin-layer, and the highly constant flow rate of the developing solvent helps the researcher to estimate the right position of the spot which can be rather difficult in the case of conventional TLC.

FF-TLC gives the possibility of the application of the sample during development of the TLC plate. This is the way of modelling the column technique by planar method, thus the behaviour of the sample components is not influenced by the α, β, γ, etc. fronts moving forward because the sample can be injected after them. Similarly, the injection of sample during development permits to avoid the complications in consequence of the double process before the separation of the sample: the sample is dried on the sorbent and solved from it by the moving mobile phase.

Another feature of FF-TLC is that the separation works as a closed layer but it is also chromatography with an open column. It means that FF-TLC can be used with taking effluent series /just as in the case of column chromatography/ and using multiple sample lines and multiple outlets: in this way, one, two or several samples can be separated in a parallel way. In addition to the possibility of parallel separation, the sorbent bed can be opened after the chromatographic process and can be checked whether any substance or component remained on the bed. If the sorbent retained one or several components, data evaluation can be performed on these remaining spots just as in the case of classical TLC, and this is an additional advantage over column chromatography separations. If all components are eluted from the thin-layer, regarding TLC with FF as a self-regenerating process, the separation /including sample application, development and elution with FF-TLC and taking effluent series/ can be continued. This procedure can be variation of the multiple useable, sintered silica gel /15/.

FF-TLC has also a good feature for displacement thin-layer chromatography /DTLC/. As it is known /16/ DTLC needs the presence of the fully developed displacement train. Although the run for the necessary length for the fully developed displacement train is short /only several cm/, because of the possible low amount of displacer in the carrier a very low R_f of the secondary /displacer/ front is achieved. With FF-TLC, the position of the secondary front can also be flexibly moved: i.e. with the overrun of the solvent /carrier/ front, the displacer front can be put into the required position. Thus, two different practical points can be reached:

512

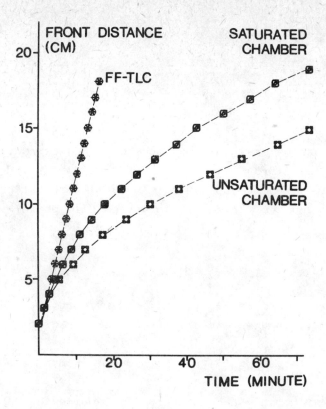

Fig. 9   Declining shape of front distance vs. time diagrams
in conventional TLC and straight line of the same
characteristics in FF-TLC.

/a/   the displacement of substances is possible, with very
low concentration of displacer,

/b/   carrying out further investigation on the theory of
DTLC, doing quantitative measurements of the parameters under
the same conditions as it has been performed in LCC. /In HPLC,
the displacement front reaches the end of the column in each
case, while in DTLC the displacer front shows different mobility and position at the end of DTLC depending on the displacer
concentration/.

SUMMARY

Forced Flow Thin-Layer Chromatography /FF-TLC/ is the separation method in which the stationary phase is arranged in planar form and is tightly covered by a membrane, and the mobile phase is supplied by a pump or a pumping device.

FF-TLC has several features including the reproducibility of TLC, extended front distance, fast development and constant flow rate as the consequences of the total elimination of the vapor phase and the forced flow of the mobile phase.

REFERENCES

/1/ Tyihák, E. and Held, Gy., in Niederwieser, A. and Pataki, G., eds., Progress in Thin-Layer Chromatography, Science Publishers, Ann Arbor, Mich, 1971. p. 183

/2/ Zakaria, M., Gonnard, M.-F. and Guiochon, G., J. Chromatogr., 271, 127 /1983/

/3/ Brenner, M. and Niederwieser, A., Experientia, 17, 237 /1961/.

/4/ Saunders, D.L. and Snyder, L.R., J. Chromatogr. Sci., 8, 706 /1970/.

/5/ Hara, S. and Mibe, K., J.Chromatogr., 66, 75 /1972/.

/6/ Soczewinski, E. and Wawrzynowicz, T., J. Chromatogr., 218, 729 /1981/.

/7/ Soczewinski, E., J. Chromatogr. 138, 443 /1977/.

/8/ Determann, H., Gelchromatographie, Springer Verlag, Berlin, 1967. p. 57.

/9/ Kalász, H., Nagy, J., Mincsovics, E. and Tyihák, E., J. Liquid Chromatogr., 3, 845 /1980/.

/10/ Tyihák, E., Kalász, H., Mincsovics, E. and Nagy, J., Proc. Hung. Ann. Meet. Biochem. 17, 183 /1977/.

/11/ Kalász, H., Tyihák, E. and Mincsovics, E., in Frigerio, A. and McCamish, M., eds., Recent Developments in Chromatography and Electrophoresis. Vol. 10. Elsevier, Amsterdam, 1980. p. 289.

/12/ Kalász, H. and Nagy, J., J. Liquid Chromatogr. <u>4</u>, 985
/1981/.

/13/ Szepesi, G., J. Chromatogr. <u>290</u>, 127 (1984)

/14/ Polyák, B., J. Liquid Chromatogr. in press.

/15/ Okumura, T., J. Chromatogr. <u>184</u>, 37 /1980/

/16/ Kalász, H. and Horváth, Cs. in Kalász H., ed., New
Approaches in Chromatography, Elsevier, Amsterdam, 1984.
p. 57.

HIGH-PERFORMANCE REUSABLE PLATES FOR THIN-LAYER CHROMATOGRAPHY AND THEIR APPLICATION TO THE ANALYSIS OF LIPIDS AND SYNTHETIC POLYMERS

E.S. GANKINA, L.S. LITVINOVA, I.I. EFIMOVA, V.E. BASKOVSKY and B.G. BELENKII

Institute of Macromolecular Compounds of the Academy of Sciences of the USSR, Leningrad, USSR

High performance thin-layer chromatography (HPTLC) has occupied an important place among the chromatographic methods. It can be used most effectively in the analysis of a large number of samples undergoing the same type of analysis with quantitative evaluation of a limited number of components, which is characteristic of most industrial analyses, chemical investigations and the analyses concerned with environment protection.

For qualitative investigations (identification of the sample components), i.e. in bioorganic chemistry and some other fields of modern chemistry, the importance of HPTLC in series analyses is even greater.

The important properties of HPTLC are as follows: it is easy to carry out adsorption and two-dimensional chromatography, the requirements for the quality of the adsorbent and the purity of the sample and the eluent are less stringent and it is possible to determine all sample components (even irreversibly adsorbed components).

HPTLC (1) is a modern development of thin-layer chromatography. As we hawe shown in 1967 (2), the use of an optimum size of adsorbent particle permits to increase the speed of analysis by a factor of about 10 and to increase the sensitivity of the determination hundred-fold (3-5).

Stabilization of the TLC conditions (use of ready-made plates and chambers with gas phase stabilization U-chambers) allows a drastic increase in the reproducibility of TLC so that it may approach that of column chromatography.

The efficiency of the analysis is greatly increased because special instruments have been designed for sample spotting (Linomat and Nanomat, Camag, Switzerland) and special procedures have been developed including thin layer chromatography under pressure (8) ("Chrompres 10", Labor MIM, Hungary), circular (1) and anticircular (9) HPTLC for the separation of a great number of samples with improved resolution at low and high R_f values, respectively.

Two-dimensional TLC permits to achieve a peak capacity of many tens of components. In HPTLC the consumption of the eluent and the adsorbent is greatly reduced. The former is decreased by using sandwich chambers and the latter by using reusable plates and reducing their dimensions.

A considerable progress has also been made in HPTLC in the development of quantitative methods in which the non-linear dependence of the detector signal on the amount of the substance in a chromatographic spot is considered and it is possible to analyze irregularly shaped zones by using multistep scanning with a point light spot scanning device (6,7).

The chromatographic plate is a major component of HPTLC. This plate consists of a support, the adsorbent and the binder.

General requirements to silica gel adsorbents for HPTLC are a particle diameter of 5-10 μm, the dispersion of particle size not exceeding 15-20%, a pore size of 100-120 Å and a pore volume of not less than 1 ml/g.

Two other components of the plates, the binder and the support, are additional and hence their properties should ensure efficient functioning of the sorbent in the chromatographic layer. These properties are chemical resistance, mechanical strength, optical properties (transparence and absence of fluorescence), no effect on the R_f value and the possibility of removing the sorbent from the chromatographic zone to extract the substance. In this sense ideal binders and supports are such substances the properties of which correspond to those of the adsorbent (silica gel). These properties are exhibited by a quartz plate used as the support and a stable silicic acid sol (silica sol) used as the binder. On heating, silica sol is transformed into silica gel with controlled porosity (10). However, quartz plates are too expensive to use in mass plate production. Moreover, for most analyses the transparence of the chromatographic layer at $\lambda > 300$ nm (the transmission range for glass) is required and glass can also be used. The chemical resistance of glass corresponds to that of silica gel. Hence, a glass support is optimal for chromatographic plates and silica sol binder is an ideal binder for plates coated with silica gel. These plates may be treated with strongly corroding agents and regenerated by a hot chromic acid mixture.

It is of interest to consider the advisability of using a silica sol binder for other supports used in the preparation of thin-layer plates. If aluminium foil is used as the support, it is also advisable to use silica sol binder because aluminium is stable to strong acids in the cold (with the exception of perchloric acid); consequently, if this binder is used, regeneration can be carried out in a chromic acid mixture at room temperature. These plates can also be reused. In the case of a polymeric support, it is not desirable to use an inorganic binder because the corrosion resistance of these plates is determined by the properties of the polymeric support.

There is another chemically resistant binder for silica gel plates. This is fusible glass used to fix the silica gel on a glass support in the so-called "sintered" plates (11,12).

These plates are chemically stable and therefore they can be regenerated in a chromic acid mixture and reused. However, "sintered" plates suffer from some drawbacks:

(a) they change the R_f of the substances when the same eluents are used as for common silica gel plates,

(b) it is impossible to remove silica gel from them for the subsequent elution of the analyses and

(c) their preparation involves complicated manufacturing processes needed for the sintering of the adsorbent layer on glass.

Today, Merck (FRG) HPTLC-plates are most widely used in HPTLC. They are prepared from silica gel with a pore size of 60 Å and a particle diameter of 5 μm. The silica gel is fixed on a glass, aluminium or polymeric support with an organic

polymeric binder (3). These plates are successfully used for
the analysis of many substance classes. However, the use of a
silica gel with medium pore diameter limits the applicability
of these HPTLC plates in the analysis of high-molecular-weight
compounds, and the presence of an organic binder prevents the
use of corroding reagents (such as strong inorganic acid and
oxidizing agents) and high temperatures in detection and plate
modification.

The aim of our investigations was to develop HPTLC plates
free of these defects. For this purpose we used a large-pore
KSKG silica gel with narrow fractions and a pore diameter of
120 Å as the adsorbent and employed a stabilized silicic acid
sol (silica sol) as the binder (14-17). This ensured not only
the possibility of using corroding reagents for detection and
plate modification but also that the plates can be regenerated
in a hot solution of a chromic acid mixture for reuse.

MATERIALS AND METHODS

KSKG silica gel (Salavat City, USSR), HPTLC-Alufolien and
DC-Alufolien (Merck, Darmstadt FRG), polystyrene standards
(Water Associates, Millford, USA), DNS- and DNP-derivatives of
amino acids (Serva, FRG) and a dye test mixture (Merck, Darmstadt, FRG) were used. The analytical grade, pure solvents,
were used without any additional purification.

GRINDING AND FRACTIONATION OF SILICA GEL

A ball mill with porcelain balls was used to grind the
silica gel, and fractionation was carried out with help of an
Alpine MZR-100 (FRG) air separator.

PREPARATION OF SILICIC ACID SOL (10)

A 114-g amount of sodium silicate (30% of SiO_2, model 2.65-3.4) was diluted to 1 liter with distilled water and the pH of the solution was adjusted to 6 by addition of wet cation exchange resin in the H-form. The resin was filtered off and the pH of the filtrate was increased to 7.4-7.8 by the addition of a NaOH solution. A 200-ml amount of the filtrate was poured into a 1 liter round-bottomed flask and refluxed for 30 min, then the remaining filtrate was added to the boiling solution from a dropping funnel for 3 h; subsequently the solution was boiled for another 30 min.

The silicic acid sol obtained may be stored for a year. Prior to use the pH of the sol was adjusted to 7 and the solution was diluted with water by a factor of two.

PLATE PREPARATION

A suspension of 4 g of KSKG silica gel (10 μm fraction) and 10 ml of silicic acid sol was stirred for 1.5-2 min with the aid of a homogenizer and poured into an applicator from a KTKh-0.2 set (Special Design Bureau of Analytical Instruments, Academy of Sciences of the USSR, Leningrad). At a slit width of ~130 μm 12-15 6x6 cm plates with a layer thickness of 130-160 μm were obtained.   The plates were dried in air at room temperature for 1 h and activated at 120°C for 30 min.

For the preparation of 10 pieces of 6 x 6 cm plates without an applicator 2 g of KSKG silica gel (10 μm fraction) and 8 ml of silica acid sol were stirred with a magnetic stirrer until a homogeneous suspension has been obtained. Subsequently, without switching off the stirrer, 0.8 ml of the suspension was withdrawn with a pipette and applied to the 6 x 6 cm plate in a uniform layer. The plates were dried with an IR lamp until the setting of the layer and then dried in air until completely dry. Prior to use the plates were activated at 120°C for 30 min.

PLATE REGENERATION

The plates were immersed for 1-2 min in a fresh chromic acid mixture heated to 150-180°C and then transferred to a glass with hot water. After cooling they were placed in a stream of tap water until complete discoloration has occurred and then repeatedly rinsed with distilled water and activated for 30 min at 120°C.

CHROMATOGRAPHY AND DETECTION

The chromatography of lipids and their detection by spraying the chromatograms with various reagents were carried out by published methods (18-23). The detection of choline-containing lipids with phosphomolybdic acid was carried out as described (24) but without additional fixing of the layer with an organic binder.

The chromatography and detection of polymers were carried out by published methods (5).

RESULTS AND DISCUSSIONS

Our HPTLC plates prepared with various particle size KSKG silica gel, with silica sol as the binder, were compared to the HPTLC-Alufolien and DC-Alufolien Merck (FRG) plates with respect to efficiency (performance), the possibility of separating synthetic polymers, and of plate regeneration as well as the possibility of their application in the analysis of lipids in which strongly corroding agents are used for detection.

The effect of particle size on the performance of the KSKG silica gel plates is shown in Fig. 1. It is clear that silica gel with a particle size of 5 and 10 μm results in high-quality chromatograms with circular spots. This fact indicates that spreading is not affected by the mass transfer between the mobile and the stationary phases (25).

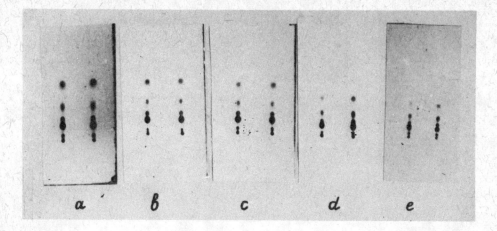

Fig. 1. HPTLC of a dye test mixture in toluene on plates with
KSKG silica gel with the following particle diameters:
a) ~ 15 μm, b) ~ 10 μm, c) ~ 5 μm and on Merck plates:
d) HPTLC-Alufolien and e) DC-Alufolien

When a silica gel with a particle diameter of 15 μm is
used, the spots become slightly elongated. The performance of
the 15 μm plates approaches that of the Merck TLC plates for
common TLC (DC-Alufolien). The efficiency of the plates with 5
and 10 μm silica gels corresponds to that of the Merck HPTLC-
Alufolien.

It should be noted that a narrow silica gel fraction is
necessary to prepare high-performance plates (26).

The efficiency of these plates was checked by the analysis
of dansyl amino acids. Fig. 2 shows the two-dimensional TLC of
dansyl-amino acids on a 6 x 6 cm plate in one of the two
solvent systems used (3). The efficiency of separation of these
acids was so high that it was possible to separate the mixture
at a distance of solvent development of only 3 cm and to obtain
four chromatograms on a 6 x 6 cm plate.

Fig. 3 shows the dependence of flow rate of these plates
for the mean particle size of the sorbent. It is clear that the
rate for the HPTLC-Alufolien is intermediate between the rates
for plates with 5 and 10 μm KSKG silica gel.

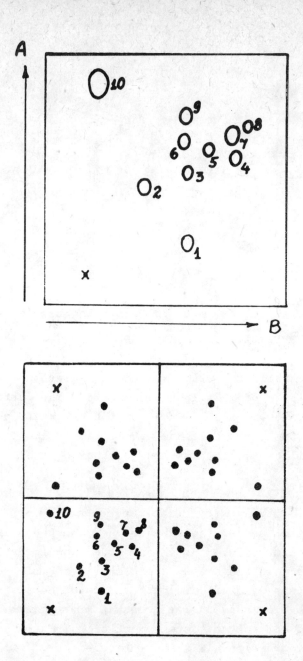

Fig. 2. Two-dimensional HPTLC of standard dansyl (DNS) amino acids.
Plates: (a) 6 x 6 cm, (b) 3 x 3 cm. Eluents: 9:7:05 acetone - isopropanol - 25% ammonia, direction $\vec{A}$ (twice); 6:4:5:0.2 chloroformbenzyl alcohol - ethyl acetate - glacial acetic acid.
Spots: 1 PRO, 2 GLY, 3 di-DNS-TYR, 4 LEU, 5 PHE, 6 ALA, 7 VAL, 8 ILE, 9 di-DNS-LYS, 10 DNS-OH(3)

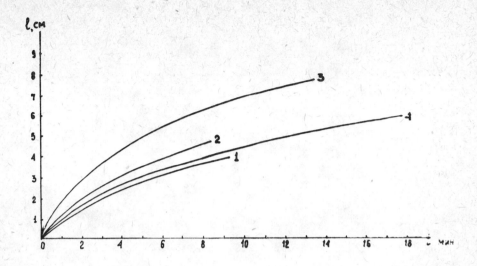

Fig. 3. Time (t) vs. distance of the toluene front migration
(1) on the following plates:
KSKG silica gel (1) 5 μm, (2) 10 μm, (3) 15 μm
particles; (4) HPTLC Alufolien Merck

Fig. 4 compares the separation of a mixture of narrow-
disperse polystyrene standards having a molecular weight (MW)
range of $20 \cdot 10^3$ to $411 \cdot 10^3$ and a pllystyrene sample with wide
molecular weight distribution (MWD) on plates coated with KSKG
silica gel and on HPTLC-Alufolien. It is clear that HPTLC-Alufolien are not very effective in the adsorption TLC of polymers.
It should be noted that the KSKG silica gel is also efficient
in the separation of low molecular weight substances; it is not
inferior in this respect to the 60 Å Merck silica gel.

One of the most interesting properties of these plates is
the possibility of their multiple regeneration in a hot chromic
acid mixture with subsequent washing with water. It is noteworthy that under these conditions silica gel is dehydrated and
acidified and its adsorption properties with respect to strongly
polar substances are modified. Fig. 5 shows that after the
plate has been regenerated by a chromic acid mixture, the adsorption activity of silica gel with respect to dinitrophenyl
(DNP) amino acids decreases drastically. However, it is completely restored when the layer is treated with a 1:1 acetone-
- 25% ammonia solution.

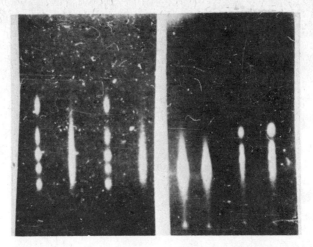

Fig. 4. HPTLC of a mixture of (1) narrow-disperse polystyrene
standards and (2) wide-molecular weight distribution
polystyrenes.
Plates: (a) coated with KSKG silica gel (pore diameter:
120 Å); (b) HPTLC Alufolien Merck prepared with silica
gel (pore diameter: 60 Å). Eluent: 17:2:1.6 cyclohexane-
toluene - methyl ethyl ketone (5). Detection was
carried out with 3.3% $KMnO_4$ in conc. H_2SO_4, with subsequent heating at 180°C for 10 min.
Molecular weights of the individual polystyrenes in the
narrow-disperse polystyrene standard mixture (from the
top to the bottom): 20,000, 51,000, 98,000, 173,000 and
411,000

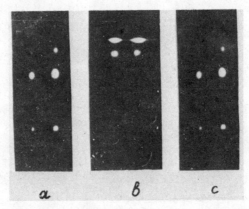

Fig. 5. HPTLC of a mixture of DNP-amino acids (from top to
bottom) PHE, ALA, SER (a) on a plate with KSKG silica
gel and silica sol and (b) on the same plate after it
has been regenerated by a chronic acid mixture, and on
plate (b) after it has been treated in a acetone - 25%
ammonia (1:1) mixture (c).
Eluent: chloroform - benzyl acohol - glacial acetic
acid (14:6:0.4) (3)

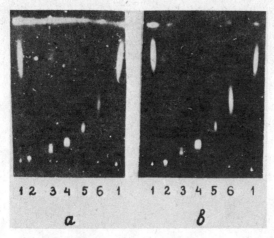

1 2 3 4 5 6 1   1 2 3 4 5 6 1

a   δ

Fig. 6. HPTLC of narrow-disperse polystyrene standards, on a
plate (a) before, and (b) after regulation in a chronic
acid mixture. Eluent: 10:9:1 CCl_4-n-heptane - methyl
ethyl ketone. Detection as in Fig. 1. Molecular weights
of the polystyrenes: 1 500,000, 2 111,000, 3 51,000,
4 33,000; 5 20,000 and 6 10,000

As to the substances with low polarity including synthetic
polymers and lipids the regeneration of the plates in the
chromatic acid mixture virtually did not affect their R_f value.
This treatment favors the burning of dust particles and organic
substances contaminating the silica gel layer; in the subsequent TLC with development by corroding agents the substances
undergoing chromatography are observed on pure white background, which is particularly important for quantitative determinations. This can be seen in Fig. 6 which shows thin-layer
chromatograms of polystyrene standards on a freshly prepared
and on a regenerated plate. It can also be seen that in the
treated plate organic impurities concentrated at the eluent
front have been burnt and this permits to detect the substances
migrating with the solvent front.

The advantages of these plates can be clearly seen if we
take as examples the analysis of lipids of marine organisms
carried out at the Institute of Sea Biology of the Far Eastern
Scientific Centre of the Academy of Sciences USSR.

Figs 7 and 8 compare the separation of neutral (one-
dimensional HPTLC) and polar (two-dimensional HPTLC) lipids

527

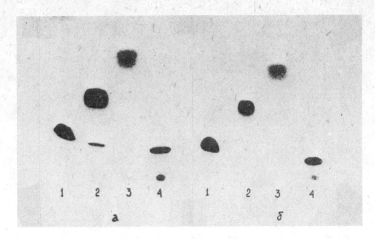

Fig. 7. HPTLC of the neutral lipids (1) cholesterine, (2)
stearic acid (3) tristearoylglycerine and distearoyl-
glycerine, (4) stearic acid methyl ester) in a hexane
- diethyl ether - acetic acid (85:15:1) eluent on a)
silica gel - gypsum and b) silica gel - silica sol
plates. Detection was carried out with a 10% H_2SO_4
solution in ethanol with subsequent heating at
180°C (4)

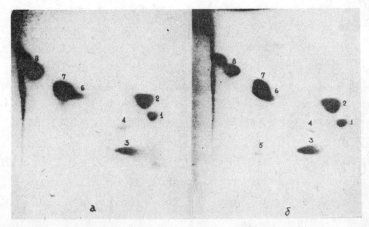

Fig. 8. Two-dimensional HPTLC of the rat brain lipids (1)
sphigomyelin (2) phosphatidylcholine, (3) phosphatidylserine, (4) phosphatide acid, (5) phosphatidylinosite,
(6) sulphatides, (7) cerebrosides in the following
solvent systems (18): chloroform - methatnol - benzene
- ammonia (65:30:10:6) - direction A, and chloroform -
- methanol - acetone - benzene - acetic acid - water
(70:30.5:10:4:1) - direction B, on plates prepared with
a) silica gel - gypsum and b) silica gel - silica sol.
Detection as Fig. 7

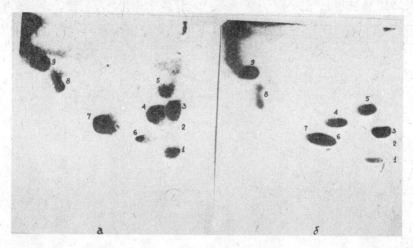

Fig. 9. Two-dimensional HPTLC of the sea grass lipids (1)
phosphatidylinosite, (2) phosphatidylserine, (3)
phosphatidylcholine, (4) phosphatidylglycerine, (5)
phosphatidylethanol amine, (6) sulphochinolyldi-
glyceride, (7) galactosyldiglyceride, (8) cerebrosides,
(9) monogalactosyldiglyceride in the following solvent
systems (19): chloroform - acetone - methanol - formic
acid - water (100:40:20:8) - direction A and acetone -
- benzene - formic acid - water (200:30:10) - direction
B on plates with a) silica gel - gypsum and b) silica
gel - silica sol. Detection as in Fig. 7

(18) on the KSKG silica gel plates with gypsum and silica sol
as the binders. It is clear that the nature of the binders does
not affect the separation of these compounds. The R_f values of
neutral lipids remained virtually invariable after 15-20 regene-
rations of the plates in a hot chromium mixture and subsequent
washing with water (complete plate regeneration took 20-30 min).
The separation of polar lipids of algae on plates with silica
sol has even improved (Fig. 9) (19). The sorbent layer with
silica sol has not been damaged by any specific and non-specific
reagent used for lipids including the reagent based on malachite green (20) which requires preliminary calcination of the
wetted plates impregnated with perchloric acid (Fig. 10) (21).
This fact makes the plates with a silica sol adhesive particularly convenient for the study of the composition of complex
lipid mixtures when the chromatogram is successively treated

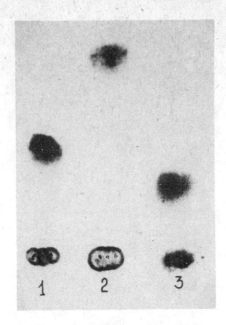

Fig. 10. HPTLC of polar phosphates. (1) glycerophosphocholine,
(2) glycerophosphate, (3) phosphocholine in the
methanol - water - ammonia (5:3:1) eluent (21).
Detection: treatment with perchloric acid and heating
at ~ 200°C with subsequent treatment with a reagent
containing malachite green (20)

with various reagents, for example, according to the following
scheme: ninhydrin is used to detect lipids containing a free
amino group, a "molybdate reagent" (22) is employed for the detection of the main phospholipids with subsequent heating at
180°C for the detection of all lipids present in the mixture,
and the reagent based on malachite green (20) is used to control
the results obtained by using a "molybdate reagent" and to
detect phospholipids present at very low concentrations.

To detect lipids on plates prepared with silica sol, procedures requiring the immersion of chromatograms into various
reagents and washing with water may also be successfully used.
For example, it is possible to use phosphomolybdic acid to
detect lipids with quaternary ammonium bases (24). These
plates are also very suitable to chromatography using sorbent
impregnation with various reagents by immersing them into an

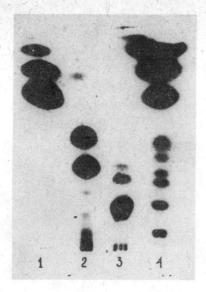

Fig. 11. HPTLC of methyl esters of fatty acids: (1) soya oil,
(2) arachidonic acid, (3) eicosapentaentic acid, (4)
noctiluca lipids on a silica gel - silica sol plate
impregnated with saturated solution of silver nitrate
in methanol in a hexane - diethyl ether - acetic acid
(96:3:3) eluent (twice) (27). Detection as in Fig. 7

appropriate solution. Fig. 11 shows the HPTLC of fatty acid
methyl esters on a silica gel layer with silica sol, with plate
impregnation with a silver nitrate solution (27).

A silica sol adhesive permits the preparation of mecha-
nically strong layers and at the same time the removal of a
silica gel containing the separated components from the plate,
without any difficulty, using a micropalette-knife, for subsequent investigation by independent methods. This procedure
employing plates with silica sol has been successfully used in
the measurements of the radioactivity of lipids and the determination of the phosphorus content of phospholipids after
silica gel has been removed from the plates, by applying the
methods developed by Vaskovsky et al. (22, 23). With this
purpose the mechanical strength of the chromatographic layer
may be controlled by using sorbent suspensions with different
silica sol concentrations for plate preparation.

CONCLUSION

High-performance thin-layer plates with a large-pore KSKG
silica gel fixed with silica sol on a glass support have been
developed. These plates exhibit a unique combination of properties making it possible to use them for the analysis of both
low and high molecular weight substances and permitting the
application of corroding reagents (such as strong acids and
oxidizing agents) and high temperatures for detection and plate
regeneration. These plates do not change the R_f of the substances being separated when the same eluents are used as for
common silica gel plates, and it is also possible to remove the
silica gel from the chromatographic zone.

Successful application of these high-performance thin-
layer reusable plates with an inorganic binder to the investigation of such complex compounds as lipids and synthetic polymers the separation and detection of which involve considerable
difficulties shows that these plates are efficient for the
HPTLC of any substance classes.

SUMMARY

High-performance reusable thin-layer plates from KSKG
large-pore (120 Å) silica gel fixed on a glass support with
silicic acid sol have been developed.

These plates permit the analysis of both low and high
molecular weight substances, the use of corroding reagents
(such as strong acids and oxidizers) and high temperatures for
detection and plate regeneration.

These plates do not change the R_f of the substances being
separated when the same eluents are used as for common silica
gel plates. Due to the controlled strength of the layer it is
possible to remove the silica gel from the chromatographic zone
for the extraction of the separated substances.

These properties of the HPTLC plates can be shown most
effectively in the analysis of lipids and synthetic polymers.

REFERENCES

1 A. Zlatkis and R.E. Kaiser, HPTLC. High. Performance Thin-Layer
   Chromatography. Amsterdam, Elsevier, 1977.
2 B.G. Belenkii, E.S. Gankina and V.V. Nesterov, Dokl. Acad.
   Nauk SSSR, 172 91 (1967).
3 B.G. Belenkii, V.V. Nesterov and E.S. Gankina, Fizicheskie i
   fizikokhimicheskie methody analiza organicheskikh
   soedinenii Nauka, Moscow, 1970, p. 80.
4 V.I. Svetashev and V.E. Vaskovsky, J. Chromatogr. 67 376 (1972).
5 B.G. Belenkii and E.S. Gankina, J. Chromatogr. (Chromatogr.
   Rev.), 141 13-90 (1977).
6 Instrumental HPTLC (ed. W. Bertsh and R.E. Kaiser) Dr. A.Huthig,
   Heidelberg, Basel-New York, 1980.
7 C. Touchstone, J. Sherma, Densitometry in Thin-Layer Chromatography. Practical and Applications. J. Wiley and Sons,
   New York 1979.
8 E. Tyihák, E. Mincsovics and H. Kalász, J. Chromatogr. 174 75
   (1979).
9 H.J. Issaq, J. Liquid Chromatogr. 4 1393 (1981).
10 V. Kh. Dobruskin, G.M. Belotserkovsky, V.F. Karelskaya and
   T.G. Plachenov, Zhur. Prikl. Khim. 40 2443 (1967).
11 T. Okumura, T. Kadano and N. Nakatani, J. Chromatogr. 74 73
   (1972).
12 T. Okumura, J. Chromatogr. 184 37 (1980).
13 Prospect Camag, Thin-Layer Chromatography, TL-8-E. Muttenz,
   Switzerland.
14 M.F. Bechtold and O.E. Snyder, Patent USA No 2574902 (1951).
15 B. Engelbrecht, H. Walker, J. Blome and F. Jordan, Patent BRD
   No 2015672 (1970).
16 E.K. Seybert, P.W. Link, U.S. Patent No 3594217 (1971).
17 J. Viska, Z. Uhrova and J. Danek, Ch. SSR Patent No 149185
   (1973).
18 V.E. Vaskovsky and T.A. Terekhova, J. High Resol. Chromatogr.
   Chromatogr. Comm. 2 671 (1979).
19 V.E. Vaskovsky and S.V. Khotimchenko, J. High Resol. Chromatogr.
   and Chromatogr. Comm. 5 635 (1982).

20 V.E. Vaskovsky and N.A. Latyshev, J. Chromatogr. 115 246 (1975)

21 N.A. Latyshev and V.A. Vaskovsky, J. High Resol. Chromatogr.
   and Chromatogr. Comm. 3 478 (1980).

22 V.E. Vaskovsky, E.Y. Kostetsky and I.M. Vasendin, J. Chromatogr
   114 129 (1975).

23 V.E. Vaskovsky and T.A. Terekhova, J. High Resol. Chromatogr.
   and Chromatogr. Comm. 2 630 (1979)

24 R.B. Schnider, J. Lipid Res. 7 169 (1966).

25 B.G. Belenkii, V.V. Nesterov, E.S. Gankina and M.M. Smirnov,
   J. Chromatogr. 31 360 (1967).

26 B.G. Belenkii, V.I. Kolegov and V.V. Nesterov, J. Chromatogr.
   107 265 (1975).

27 P.A. Dudley and R.E. Anderson, Lipids, 10 113 (1975).

Chromatography, the State of the Art
H. Kalász and L.S. Ettre (Eds)

THIN-LAYER ION-EXCHANGE CHROMATOGRAPHY IN BIOCHEMICAL ANALYSIS

TIBOR DÉVÉNYI and HUBA KALÁSZ*

Institute of Enzymology, Biological Research Center
of the Hungarian Academy of Sciences, Budapest, Hungary
*Department of Pharmacology, Semmelweis University of
Medicine, Budapest, Hungary

INTRODUCTION

Thin-layer chromatography (TLC) has been developed to a considerable extent during the past few years. Major advances have been encountered in the choice of adsorbents (introduction of high-performance TLC, reverse-phase TLC, the use of sintered TLC plates, etc.) in the general techniques (introduction of the U-chamber, circular and anticircular techniques, over-pressured TLC, linear-sandwich TLC, etc.) and, last but not least, in the quantitative evaluation techniques (new detectors, e.g. Vidicon-tubes, microprocessor control, etc.). A good survey of these developments including references of 54 papers is found in a review of Issaq (1981).

Many different ion exchangers have been used mainly for the separation of closely related compounds, although silica gel remained the major solid phase in TLC. In the early 1950s home-impregnated filter papers were used, followed by the introduction of commercially available products. A few years later as TLC was substituted for paper chromatography for well known reasons, various forms of thin-layer ion-exchange techniques were described. In addition to the home-made ion-exchanger layers some precoated ion-exchanger chromatoplates and sheets became commercially available and, as pointed out by Sherma and Fried (1982), were widely accepted. A review was published by Elődi and Karsai (1980) on the application of thin-layer ion-exchange chromatography in biochemistry and related fields.

535

GENERAL ASPECTS

1. Ion-exchangers in TLC

Thin-layer ion-exchange chromatographic techniques can be performed either by impregnation of a conventional TLC plate or sheet with an ionic compound or by application of a TLC layer prepared with an ion exchanger.

A. Impregnation methods in TLC

Silica gel plates or sheets can be converted to ion-exchange layers with various ionic compounds. Most of the papers published on the application of impregnated silica gel TLC plates or sheets reported on the separation of ionorganic ions only. Various oxides, hydrous oxides, organic resins, liquid ion-exchangers, ammonium molybdenum phosphate, ammonium tungsten phosphate, zirconium phosphate and other similar compounds were used as impregnating media. These layers permitted the application of aqueous solutions containing inorganic salts. This subject was reviewed in detail by Fuller (1971).

Yasuda (1973) employed a metal salt for the impregnation of silica gel for the separation of aromatic amines. On the other hand, Srivastava et al. (1979) used various anions for the impregnation of silica gel in the study of the separation of closely related aromatic amines. Both authors used ionic media as the stationary phase while the eluents were organic solvents (benzene and ethanol); consequently it is probable that the separation is due to a complex retardation instead of a pure ion-exchange reaction.

In the recent literature the impregnation of more or less indifferent layers (such as silica gel) plays only a minor role, however, it may have great importance in other fields (e.g. reversed-phase TLC). Lepri et al. (1981) described a TLC method using C_2, C_8 and C_{18} reversed-phase plates for the separation of amino acids in biological fluids.

B. Solid ion-exchangers in TLC

A large list of ion-exchangers used for TLC was compiled
by Stahl and Mangold (1975) and by Walton (1975) in the handbook for chromatography edited by Heftmann (1975). This book
also includes the list of suppliers.

Home-made ion-exchange chromatoplates were mainly prepared
from a mixture of cellulose and any, either weak or strong,
synthetic ion-exchange resins. A mixture of cellulose and Bio-
Rad AG1-X4 was used in acetate form for the separation of
alkaloids (Lepri et al. 1976a).Various layers as e.g.TLC layers
containing alginic acid, strongly acidic Dowex-type resin, or
CM-cellulose, mixed with microcrystalline cellulose were also
studied (Lepri et al. 1976b). The chromatographic behaviour of
anionic layers in the separation of phenolic compounds was also
investigated (Lepri et al. 1976c). Muchova and Joke (1978)
reported on mixed TLC layers prepared from a mixture of
Amberlite-IRA-400 and silica gel. The three-component system
composed of CM-cellulose, silica gel and starch was examined by
Gocan (1980). Senyavin et al. (1980) determined ion-exchange
constants for mixed layers using ion-exchange TLC. It is noteworthy that mixed layers can also be prepared from non-ionic
resins, like Amberlite XAD for the study of adsorption on
resin-coated TLC. These layers were prepared with a binder such
as $CaSO_4$ or silica gel, glass powder poly(vinyl alcohol),
cellulose, etc.(Pietrzyk et al. 1979).

The ready-for-use precoated ion-exchange chromatosheets or
plates are easier to handle. Among them the most widely used
are the cellulose derivatives, as CM-, DEAE- or PEI-cellulose
TLC plates or sheets. Judging after the number of references,
PEI-cellulose ion-exchange chromatoplate appears to be the most
often used material which is widely employed in the analytical
biochemistry of nucleic acids. This subject was first reviewed
by the pioneers of this analytical tool: Randerath and
Randerath (1967), later by Pataki (1968) and by Böhm and
Schultz (1974).

Strongly acidic and basic resin-coated chromatosheets are
commercially available. Fixion 50X8 in Na^+ form is the cation

exchanger and Fixion 2X8 in acetate form is the anion exchanger. The manufacturers and suppliers are Chromatronix (Mt. View, Calif. USA) and Medimpex (Budapest, Hungary). In the Federal Republic of Germany Macherey-Nagel (Düren) is producing these resin-coated chromatosheets with the trade name Ionex 25-SA for cation exchanger and Ionex 25-SB for the anion exchanger. These resin-coated chromatosheets (on a polyethylene terephtalic ester plastic) are prepared from a mixture of Dowex type resin, silica gel and an indifferent organic binder as a minor component. As pointed out by Lederer (1977) the silica gel present in the layer does not interfere with the ion exchange, except for a very few cases. In general, these layers, especially the cationic type, can be considered as analogues of the resins used in column chromatography.

It is evident that since the pioneering work of Moore and Stein (1951) ion-exchange chromatography has become a powerful method for the separation of closely related compounds. On the other hand, it is also without any doubt that since the pioneering work of Kirchner (1951) and Stahl (1956) TLC - with its incomparably elegant technical simplicity - is one of the most widely employed techniques in analytical biochemistry. TLC is highly efficient, relatively inexpensive and it permits the simultaneous handling of a large number of samples. Consequently, the combination of both methods may even provide the possibility of large-scale operations in many fields of analytical biochemistry, chemistry and medicine.

2. Pretreatment and Development

The buffers usually encountered in ion-exchange column chromatography are suitable for TLC methods as well. Deionized water is essential and the use of reagent-grade chemicals is strongly advocated since the separations are influenced by pH, molarity, and in some cases, the eluting temperature.

Non-ionic additives may improve the separation of closely related compounds. Glycerol, ethylene glycol, alcohols, methyl cellosolve, detergents may lower che tension of buffers; consequently, the diffusion of the spots at elevated temperatures may be decreased. At higher concentration of the organic solvents, adsorption or partition will substitute the ion-exchange process.

The ion-exchange resin employed for column or TLC technique must be in the appropriate ionic form. The preparation of an ion-exchange TLC plate in the laboratory begins with the equilibration of the components before the coating process. In most cases (e.g. gel filtration by TLC) an equilibration is also necessary after coating.

The equilibration of chromatoplates (i.e. thin-layers coated onto glass plates) can be performed in the following way:

A piece of a thick filter paper (15 x 20 cm) is placed around the upper edge of the chromatoplate in such a way that the paper covers a 1-cm band of the upper part of the layer. The filter paper is held in place by a 20 cm long thin glass rod and a plastic clothespin. The solvent for equilibration is filled into the chromatographic chamber and the chromatoplate is placed in it. The equilibration is carried out by a con-tinuous ascending development of the buffer. If a flexible chromatosheet is used, it should be placed with the layer facing outward on a clean dry glass plate having the same size as the chromatosheet (20 x 20 cm).

If a great number of flexible chromatosheets have to be handled simultaneously, chromatosheet holders ("Uniformizer"), as shown in the Figure 1, permit the processing of several hundred samples per day (Sigurbjörsson et al. 1979).

Diluted buffers are recommended for rquilibration. A 10-30 fold diluted developing buffer is most frequently used for equilibration. The concentration of the equilibrating solution is not critical in the case of volatile solvents (HCl, acetic acid, pyridine acetate buffers, etc.), since its excess can be eliminated by drying the chromatosheets or plates.

Commercially available thin-layer ion-exchange plates or sheets usually do not require equilibration. The latter is only necessary if:

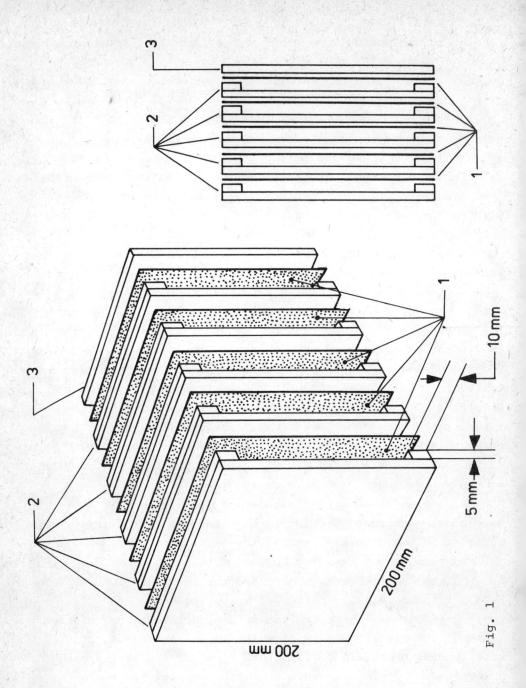

Fig. 1

540

a) closely related compounds of high R_f values (over 0.7) have to be separated; and/or

b) the ionic form has to be changed.

3. Application of the Sample

The application of the sample in the case of thin-layer ion-exchange techniques can be carried out with the same tools - glass capillaries, micro syringes or automated spotting devices - which are applied in conventional TLC methods. Regarding the size and shape or the amount of the samples, there is no difference between the two methodologies. However, there is a remarkable difference according to the preparation of the samples to be spotted.

In the case of conventional TLC techniques (mainly using silica gel as the stationary phase) only salt-free samples can

Fig. 1. Chromatosheet holder ("Uniformizer") for thin-layer ion-exchange chromatographic sheets. Plates of 20x20 cm and strips of 1x20 cm are cut out from a transparent plastic sheet having a thickness not greater than 0.5 cm. Two strips are fixed to the opposite edges of the plastic plate with a synthetic adhesive which does not swell in aqueous solutions. A few of the plates should stay without strips.

Use of the sheet-holder: put a plain plastic sheet on the table and place the first ion exchanger TLC sheet on the top of it with the resin facing upwards. Then place an "Uniformizer" plate on the resin-layer in such a way that the two strips are placed at the bottom and the top edges of the sheet, respectively. Thus each sheet is now closed in a place the height of which is determined by the thickness of the plastic strips. Now the second TLC-sheet is similarly placed onto the back of the "Uniformizer". A second sheet-holder is placed on this, which is then covered by another "Uniformizer", etc. At last the "sandwich" of plastic sheets and "Uniformizers" is fastened together with a rubber band. In such a way approximatively eight ion-exchange chromatosheets can be placed in a chromatographic tank of conventional size. After development of the chromatograms, the "sandwich" is placed onto a flat surface, the rubber band is removed and the sheets are put one by one onto a suitable horizontal holder for drying.

be applied. Thus if the sample contains a salt, e.g. as a result of neutralization, it must be desalted. This requires a rather time-consuming procedure, and moreover, it can cause a loss of the material. However, in the case of ion-exchange chromatography, the salt content applied with the sample does not affect the separation process, as the eluting (developing) buffer usually employed for the separation may have a much higher salt concentration than that of the sample. The detection and determination of tryptophan from neutralized alkaline hydrolysates can be cited as a good example. In fact in this case, the salt present in the sample does not influence the separation since the eluting (developing) buffer has 1.7 mol/l Na^+ concentration (Dévényi et al. 1971). The sample can be spotted directly onto the resin-coated chromatosheets, without preliminary neutralization if the alkaline hydrolysis of the samples has been carried out with $Ba(OH)_2$ since the salt present in the solution sticks at the origin and does not affect the determination of tryptophan (Pongor et al. 1978).

Acidic hydrolysates can also be applied directly to resin-coated chromatosheets, without removing the acid, since the excess of the acid is instantly evaporated by hot air. In the course of a large-scale screening program (Sigurbjörnsson et al. 1979) the samples to be tested for lysine content were hydrolysed in 6N HCl (400 mg wheat or barley per 2 ml 6N HCl) permitting the application of a single spot (2-3 µl) with disposable capillaries from each sample. Nearly 1000 samples were handled daily in this way.

Only special thin-layer ion-exchange layers can be used for the separation of proteins (DEAE-cellulose, CM-cellulose, etc.). Strong acidic or basic layers are only recommended for small molecules. The resolution power of the resin incorporated in the TLC chromatoplate is markedly decreased by the presence of traces of substances having large molecules, like proteins. In the practice of clinical biochemistry only protein-free samples can be applied on resin-coated chromatoplates or sheets. Several methods are known for the deproteinization of biologi cal fluids (e.g. trichloroacetic acid, trifluoroacetic acid) or for the extraction of blood (or any other biological fluid)

samples dried on filter paper. In both cases the protein-free
solution containing TCA, TFA or alcoholic HCl (used for the
elution of samples dried on filter paper) can be applied directly onto the resin-coated chromatoplates or sheets (Issaq and
Dévényi 1981, Dévényi 1982).

4. Visualization and Quantitative Evaluation of Thin-layer Ion-exchange Chromatograms

Most of the commercially available, ready-for-use ion-
exchange chromatoplates and sheets are sufficiently stable so
that they can be handled like conventional silica gel plates as
far as visualization or quantitative evaluation are concerned.
However, the consistency of the more or less unstable home-made
mixed layers prepared from cellulose and resin mixtures, etc.
must be taken into further considerations.

The visualization techniques usually applied in TLC,
namely
- photometric UV detection methods (at 254 - 366 nm), or
- photometry at visible light after color reactions
are equally suitable for the evaluation of the ready-made ion-
exchange plates and sheets either containing cellulose derivatives or manufactured from a synthetic resin mixed with silica
gel.

A long list of recipes for the visualization of different
compounds was compiled in Stahl's monograph (1965). Many of
these reactions are routinely applied for a wide range of
organic compounds.

Densitometers based on the measurement of either transmittance or reflectance, or fluorescence were designed for the
quantitative evaluation of thin-layer ion-exchange chromatograms. In the case of labelled compounds a readioscanner can be
used. A measuring method based on the video-technique, termed
"video-densitometry", and an instrument was developed for the
rapid scanning of ninhydrin-reacted chromatograms, considering
the great efficiency of thin-layer ion-exchange chromatography
on resin-coated chromatoplates with respect to amino acid analy-

sis (Kerényi et al. 1976, Dévényi 1976a, b, Pongor et al. 1978a, Pongor 1982).

TV-type multichannel detectors reveal several properties unusual in commercially available densitometers (Talmi 1975). The main difference originates from the practice of video-scanning which is very rapid and its geometry differs from that used in conventional instruments. Using a TV-type detector, the image of the chromatographic spot is projected to a target plate which is a two-dimensional array of unit detectors continuously scanned by an electron beam. These unit detectors behave like capacitors, whose charge is proportional to the incident light intensity, and which are periodically discharged by the electron beam. Optical magnification can be selected in a wide range according to the size of the spots on the chromatogram. For the purposes of mathematical treatment, the scanning scheme of video-densitometry can be considered as step-wise two-dimensional scanning (Pongor 1982).

Spectral sensitivity of Vidicon-type detectors allows quantitative determination in the visible range. Extending the range to UV light for direct measurements would require a special face plate and camera optics.

The detection limit for amino acids after visualization with ninhydrin-cadmium reagent is in the range of 2-5 nanomoles measured with a commercially available Vidicon-tube.

The output signals of the TV-camera are too fast to be integrated by conventional integrators used for liquid- or gas chromatography. With a TV-type detector system, combined with a high-speed integrator and data-processing unit, the time necessary for a measurement can be reduced drastically as compared to conventional instruments.

The idea of building a Vidicon based densitometer was first described by Hannig and Wirth (1968). The first commercial video-densitometer was available only in 1976 from Chinoin-Nagytétény, Budapest, Hungary (Kerényi et al. 1976). This instrument works according to a quasi double-beam principle, since the background intensity is sampled in each line of the scanning. The instrument can be used in either the re - flectance or transmittance mode and it is equipped with a UV-light source for fluorescence measurements. The control unit of the instrument fulfils two functions: It contains the high-speed integrator working according to a digital procedure (Kerényi et al. 1976) that can be regarded as a three-dimensional extension of the area measurement used in image analysis. The integrator performs two measurements simultaneously: one of which is of the spot in question, the other can be of the total (stained) material present on the chromatogram (or an internal standard spot). Another function is the selection of the measuring area. This is carried out by the operator with the aid of a monitor. Geometrically uniform chromatograms can be screened automatically. The measured data are displayed on the monitor and also transferred to an on-line connected programmable calculator (Hewlett-Packard HP 97). The calculator can be programmed to carry out some calculations connected directly with the measurement (calibration curve fitting, use of standards, calculation of percentage, etc.) (Pongor et al. 1978a). This method was adapted to large-scale routine work in the determination of essential amino acids in plant proteins and in many fields of clinical biochemistry.

SEPARATION OF AMINO ACIDS BY THIN-LAYER ION-EXCHANGE CHROMATOGRAPHY

1. One dimensional Separation of Amino Acids and Related Compounds

Contrary to conventional TLC methods, thin-layer ion-exchange chromatography permits the direct application of samples containing salts or acids, and the separation of amino acids can be achieved by one-dimensional developments.

In most cases room temperature is recommended for the thin-layer ion-exchange chromatographic separation of closely related compounds, though some authors suggested lower or higher eluting temperature as well. If low temperature ($+4^{\circ}$C) is needed, the chromatographic tank should be kept in a re - frigerator, while if it is essential to work above room temperature (e.g. the separation of methionine, isoleucine and

leucine can be carried out only between 37 and 45°C), the jar
should be placed in a thermostat ensuring the appropriate temperature. In such cases the tank (containing the eluting buffer)
should be equilibrated at the working temperature for 30 minutes.

Various chambers were designed for special tasks. A pres -
surized ultramicro chamber can be applied for the fast separation of amino acids on strongly acidic resin-coated chromatosheets, according to Kalász et al. (1980), Kalász and Nagy
(1981) and Cong et al. (1982). The latter method - called overpressured TLC - permits markedly shortening the elution time
while the diffusion of the spots is also decreased. This method
can be applied for the modelling of column chromatographic
separations on thin-layer plates.

A ninhydrin spray reagent can be applied for the detection
of amino acids on precoated ion exchange sheets.

A. One-dimensional Separation of Amino Acids

The separation of amino acids on resin-coated chromatosheets depends on the ionic form of the-resin (present in the
layer), and on the characteristics of the developing buffer
(pH, molarity), as well as on the eluting temperature.

Fifteen amino acids out of 16 (usually present in an acid
hydrolysate from peptides or proteins) can be separated on
strongly acidic resin-coated chromatosheets (Fixion 50X8 in
Na^+ form), applying a sodium citrate buffer pH = 3.3 (Na^+ = 0.5
mol/l, citrate = 0.4 mol/l) at 45°C (Dévényi et al. 1971b,
Dévényi 1982). Only threonine and serine form a common spot
(Fig. 2). A rather close analogy can be established between the
one-dimensional chromatogram of amino acids at pH = 3.3 and the
pattern obtained with the amino acid analyzer operating in a
single-column procedure. Only a few changes can be observed in
the order of amino acids, due to the pH differences and partly
due to the presence of silica gel in the resin-coated chromatosheet (proline and histidine have quite different positions).

The one-dimensional method for the separation of amino
acids is sensitive to slight changes in pH and in molarity. The

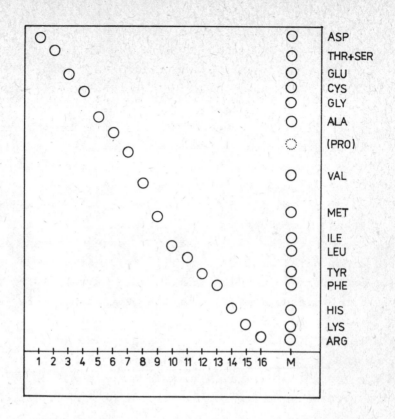

Fig. 2. One-dimensional separation of amino acids on strongly
acidic resin-coated chromatosheet (Fixion 13 50X8 in
Na^+-form). Developing buffer: pH = 3.3, Na^+ = 0.5 mol/1,
citrate = 0.4 mol/1. Ingredients per liter: citric acid
monohydrate = 83.0 g, sodium hydroxide = 16.0 g, sodium
chloride = 5.0 g, 37% hydrochloric acid (sp. gr. 1.19) =
= approx. 4.0 ml ethylene-glycol = 54.0 ml. The exact
pH should be adjusted with a pH-meter! Equilibration:
continuous development for 24 hours at room temperature
with a 30 fold diluted developing buffer. Chromatography at 45 C°. Staining with ninhydrin-cadmium-
collidine reagent.
1 = ASP, 2 = THR+SER, 3 = GLU, 4 = CYS, 5 = GLY, 6 =
ALA, 7 = PRO, 8 = VAL, 9 = MET, 10 = ILE, 11 = LEU, 12 =
TYR, 13 = PHE, 14 = HIS, 15 = LYS, 16 = ARG, M =
mixture of 1-16
Comments:
a) If the solvent front is uneven or the amino acids
   with high R_f values converge, the equilibration of the
   the resin-coated chromatosheet has been incomplete.
b) If the R_f value of ASP is less than 0.8, the pH of
   the buffer may be lower than 3.3.
c) If the separation of MET/ILE/LEU is incomplete, the
   temperature is below 37°C (as minimum for the separation of these three amino acids).
d) If the separation of PHE and HIS is incomplete, the
   molarity of Na^+ may be over 0.5.

547

separation of Met/Ile/Leu is affected by changes in the eluting temperature.

The method was adapted by Sajgó and Dévényi (1971) for the examination of peptide hydrolysates obtained with exopeptidases (e.g. carbopeptidase A+B or leucine aminopeptidase, respectively). Several authors applied these simple analytical studies of amino acid sequence micro-method in the investigations of peptides and proteins (Biszku et al. 1973, Welling et al. 1974, Sajgó and Hajós 1974, Telegdi et al. 1973, Fábián 1977, Markovic and Sajgó 1977). Clarke et al. (1976) used the method for monitoring glutathione levels. Jones et al. (1974) employed the method for the determination of amino acids in aquatic filter folders. Kisfaludy et al. (1972) studied the enzymatic cleavage of peptides containing amino-oxycarboxylic acids.

An increase in the pH of the developing buffer will cause an increase in the R_f values of the amino acids, improving the separation of the basic and aromatic amino acids which have the lowest R_f values at pH = 3.3. However, the upper part of the chromatogram deteriorates to a large extent as a consequence of the increase in the pH(=increase in the R_f values).

B. One-dimensional Separation of Basic and Aromatic Amino Acids

The separation and determination of basic amino acids is carried out on the shorter column of a conventional amino acid analyzer operated with a two-column system. The eluting buffer for the shorter column is usually sodium citrate (pH = 5.28; $[Na^+] = 0.35$ mol/l). Almost identical experimental conditions can be applied in the case of strongly acidic resin-coated chromatosheets. However, in this case, in addition to the basic amino acids, tyrosine and phenylalanine can also be separated (Dévényi 1970a, 1976). The composition of the buffer and the relative positions of the amino acids are shown in Fig. 3.

Contrary to chromatography at pH = 3.3, prewash or equilibration of the chromatosheets is not necessary for the separation of basic and aromatic amino acids: the development can be performed at room temperature, and the separation is not sensi-

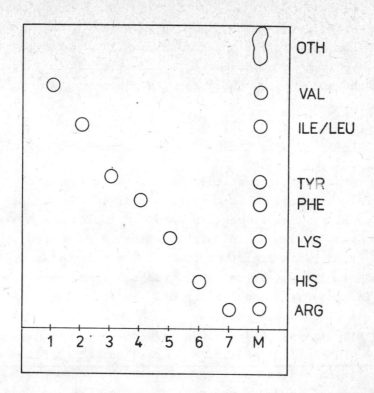

Fig. 3. One-dimensional separation of basic and aromatic amino
acids on strongly acidic resin-coated chromatosheets
(Fixion 50X8 in Na⁺-form). Developing buffer: pH = 5.1,
Na⁺ = 0.44 mol/l, citrate = 0.07 mol/l. Ingredients per
liter: citric acid monohydrate = 14.1 g, sodium hydroxide = 8.0 g, sodium chloride = 14.0 g, 37% hydrochloric acid (sp.gr. = 1.19) = 4.4 ml. Equilibration is
not needed. Chromatography at room temperature. Staining with ninhydrin-cadmium sprayreagent.
1 = VAL, 2 = ILE + LEU, 3 = TYR, 4 = PHE, 5 = LYS,
6 = HIS, 7 =ARG, M = mixture of 1-7.

tive to slight changes in the pH or molarity. This method can
be employed in many fields of analytical biochemistry including
screening of amino acid disorders, large-scale screening of
lysine content in plant breeding programs, etc.

Lysine and ornithine cannot be separated at a pH of around
5. A lower pH is required for the separation of these two amino
acids. Sharp separation can be obtained at pH = 4.2 with a
sodium citrate buffer (Hrabák and Ferenczy 1971, Issaq and
Dévényi 1981).

A good separation of hydroxy lysine can be obtained with a sodium citrate buffer at a pH of 5.23 on strongly acidic resin-coated chromatosheets (Buzás et al. 1980). The procedure is simple and well suited for the characterization of collagen type proteins.

C. Separation and Determination of Diaminopimelic Acid (DAPA)

DAPA is the constituent of the cell wall of bacteria and it is considered to be a suitable marker of microbial protein synthesis in the alimentary canal of ruminants. A simple method for the separation and determination of DAPA from crude extracts of rumen digest may have a great importance. A one-dimensional separation method was described by Pongor and Baitner (1981) using Fixion 50X8 (Na^+) strongly acidic chromatosheets at room temperature, with a sodium citrate buffer (Na^+ = 0.4 M, citrate = 0.4 M, pH = 3.14).

The sample preparation is carried out as follows:

Rumen content is strained through cheese-cloth. The liquor is allowed to sediment for 10 minutes and the supernatant is centrifuged at 7000 g for 10 minutes. 500 mg of the wet pellet is hydrolysed with 2 ml 6 N HCl and 0.5 ml 37% HCl at $105^\circ C$ for 48 hours. The hydrolysates are filtered through paper, rinsed with 2 ml of water and the combined filtrates are evaporated to dryness and redissolved in 0.2 ml water. Aliquots of 3-6 µl are spotted on the chromatosheets (corresponding to 1-4 µg of DAPA).

Methionine and isoleucine form a common spot, while DAPA is well separated between Met/Ile and Val, when a sodium citrate buffer is employed at a low pH.

D. Racemization Test for Peptide Synthesis

The detection of racemization and the estimation of it's extent is decisive for the results of peptide synthesis. Japanese authors have elaborated a simple method for the detection of racemization. The principle of this method is the following: L-Gly. Ala dipeptide is combined with L-Leu under

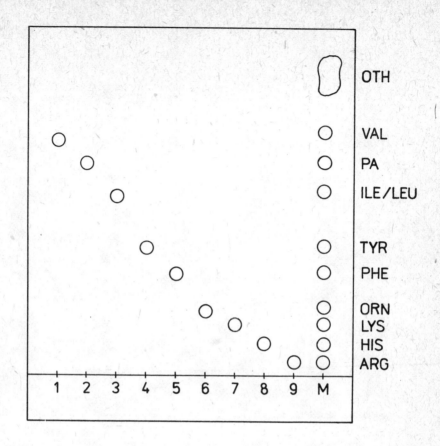

Fig. 4. Separation of D-penicillamine on strongly acidic resin-
coated chromatosheet (Fixion 50X8 in Na$^+$-form). Developing at +4°C. Buffer: pH = 4.4, Na$^+$ = 0.3 mol/l,
citrate = 0.07 mol/l. Ingredients per liter: citric
acid monohydrate = 14.1 g, sodium hydroxyde = 8.0 g,
sodium chloride = 5.85 g, 37% hydrochloric acid (sp.
gr. = 1.19) = approx. 5 ml. The exact pH should be
installed using a pH-meter. Equilibration of the
chromatosheet: continuous development for 24 hours at
room temperature with a 30-fold diluted developing
buffer. Staining with ninhydrin-cadmium spray-reagent.

1 = VAL, 2 = D-penicillamine (PA), 3 = ILE-LEU,
4 = TYR, 5 = PHE, 6 = ORN, 7 = LYS, 8 = HIS, 9 = ARG,
M = mixture of 1-9, OTH = other amino acids.

experimental conditions identical to those employed in the
given peptide synthesis. In the optimum case, the product is
exclusively L-Gly.Ala.Leu. However, D-L-Gly.Ala.Leu tripeptide
is also found in the mixture if racemization takes place under
the prevailing experimental conditions (Izumya and Muraoka

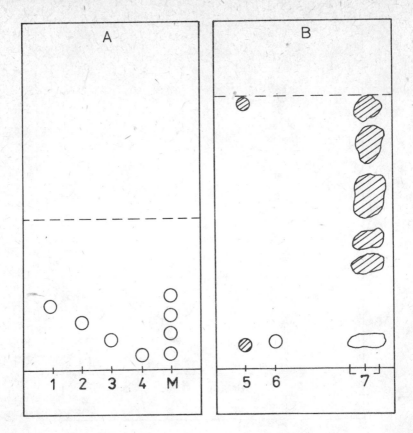

Fig. 5. Detection of tryptophan on strongly acidic resin-coated
chromatosheets (Fixion 50X8 in Na$^+$-form). Elution buffer: pH = 6.0, Na$^+$ = 1.6 mol/1, citrate = 0.04 mol/1.
Ingredients per liter: citric acid monohydrate = 7.0 g,
sodium hydroxide = 4.0 g, sodium chloride = 87.75 g,
methyl-cellosolve = 100 ml. Development at room temperature. Chromatosheets are used without equilibration
or prewash. Solvent front height in the case of free
amino acid mixtures (Tracing A): 10 cm. Front height in
the case of alkaline hydrolysates (Tracing B): 15-18
cm. 1 = LYS + HIS, 2 = PHE, 3 = ARG, 4 = TRP, M = mixture of 1-4, 5 = alkaline hydrolysate of a GLY.TRP
pipeptide sample, 6 = alkaline treated TRP, 7 = alkaline hydrolysate of a wheat sample.

Sample preparation:

a) Peptides, proteins: Approximatively 0.5 mg of the
sample is dissolved in 0.2 ml 4 mol/1 NaOH, flushed
with nitrogen, using a thin glass capillary. Hydrolysis is carried out at 105°C for 4 hours after
closing the vial(s). After cooling 0.3 ml of sodium
citrate buffer (pH = 3.28, Na$^+$ = 0.2 mol/1) and 0.1
ml 37% HCl (sp. gr. 1.19) is added to the sample
Aliquots should be spotted directly onto the
chromatosheets.

⟶

1969). The authors applied the amino acid analyzer technique for the detection and the estimation of the extent of racemization.

These two closely related tripeptides can be easily separated on strongly acidic resin-coated chromatosheets (Fixion 50X8 in Na$^+$ form). The relative distance and the experimental conditions are shown in Fig. 4. Good separation is obtained when a sample having a concentration of 10 µmol/µl is dissolved in 0.01 mol/l HCl. The samples should be applied in 1-cm bands. The chromatosheets are developed up to a sonvent-front height of 15 cm at room temperature (Dévényi 1970b).

Any of the densitometric techniques is suitable, in order to determine quantitatively the extent of racemization within an accuracy of 5-10%.

E. One-dimensional Separation of D-Penicillamine

A slight change of the pH and molarity of the developing buffer applied for the one-dimensional separation of amino acids and a rather large change of the eluting temperature permits the one-dimensional separation and determination of D-penicillamine, a thiol derivative of valine. Its therapeutic use dates back to the late 1950s. More recently it was found to be effective in the treatment of hyperbilirubinaemia of premature and mature infants. D-penicillamine can be separated from the free amino acids present in the blood sample by ion-exchange TLC employing strongly acidic resin-coated chromatosheets (Fig. 5). Sodium citrate at pH 4.4 (Na$^+$ 0.3 mol/1, citrate 0.07 mol/1) is the developing buffer (Pongor et al. 1978c, Kiss and Kovács 1982). D-Penicillamine (mainly in disulfide form) appears on the chromatogram as a well separated, single spot between valine and methionine/isoleucine/leucine. The sample preparation from blood has been described previously.

←——————

b) Plant seeds, fodders, etc.: Approx. 400 mg of the finely grinded samples are hydrolysed for 24 hours at 105°C in vials or sealed tubes with 1.2 g of Ba(OH)$_2$ in 1.5 ml deionized water in nitrogen atmosphere. Aliquots should be spotted directly after cooling without neutralization.

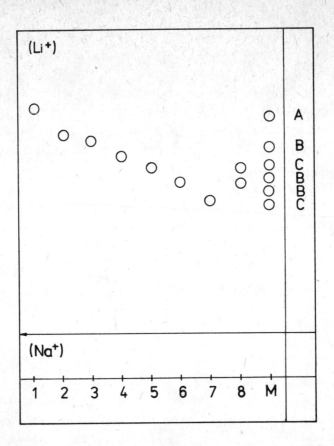

Fig. 6. Separation of amino acid mixtures containing asparagine and glutamine on strongly acidic resin-coated chromato-sheets (Fixion 50X8 in Na^+-Li^+ "tandem"-ionic form). Developing buffer: pH = 2.80 ± 0.02, Li^+ = mol/l, citrate = 0.05 mol/l, formate = 0.05 mol/l. Ingredients per liter: lithium citrate tetrahydrate = 14.10 g, lithium chloride = 2.12 g, 85% formic acid = 2.30 ml, 37% HCl (sp. gr. 1.19) = 8.0 ml. Chromatography at 37°C. Solvent front height = 20 cm. Staining with ninhydrin-cadmium-collidine reagent.

Pre-treatment of the chromatosheets:

a) The chromatosheets in Na^+-form should be converted into Li^+-form by continuous development with 1.0 mol/l lithium chloride solution for 24 hours at room temperature.

b) After drying the sheets with hot air, the excess of lithium chloride should be removed by continuous development with deionized water for 24 hours, at room temperature.

c) To convert the lower part of the chromatosheet back to the Na^+-form, it should be developed with a sodium chloride solution (1.0 mol/l) to the height of 2.5 cm above the starting line.

⟶

F. Detection and Determination of Tryptophan by Thin-layer Ion-exchange Chromatography

Triptophane (Trp) cannot be detected in acidic hydrolysates since it decomposes in this medium. In principle, two possibilities are known for the estimation of Trp: a) a specific color reaction of the alkaline hydrolysate, or b) spectrophotometric analysis of the non-destructed material. Both of these methods have several disadvantages. The color reaction of the alkaline hydrolysates is not quite specific since they may contain several interfering contamination. Spectrophotometric methods are only applicable when the substance yields a clear solution and no other components, e.g. coenzymes, etc. are present which may adsorb in the given wavelength range.

Conventional TLC of an alkaline hydrolysate is not feasible, because of the very large salt content of the solution after neutralization.

A rather good separation of Trp can be achieved by thin-layer ion-exchange chromatography on strongly acidic resin-coated chromatosheets (Fixion 50X8 in Na^+ form) employing a sodium citrate buffer (pH = 6.0) containing a large amount of sodium chloride (Fig. 6). The R_f value of Trp is the lowest among the amino acids present in an alkaline hydrolysate (Dévényi et al. 1971a, Pongor et al. 1976). Since the developing buffer contains a large amount of salt, the latter present in the neutralized sample does not interfere with the separation.

1 = ASP, 2 = THR, 3 = SER, 4 = ASN, 5 = GLN, 6 = GLU, 7 = - GLY, 8 = GLN + GLU, M = mixture of 1-8. A = bluish gray, B = purple, C = yellow

G. Separation of Amino Acid Mixture containing Asparagin and Glutamin

The separation and determination of asparagin (Asn) and glutamine (Gln) from other amino acids is of great importance in sequence analysis, clinical biochemistry, peptide chemistry, etc. Váradi (1975) developed a one-dimensional separation employing strongly acidic resin-coated chromatosheets in Na^+-Li^+ "tandem" ionic form (Fig. 7). The special ionic form is prepared from the commercially available sheet in Na^+ form, applying the equilibration steps described in the legend of Fig. 7. As a result of this procedure the large (upper) part of the chromatosheet is in the Li^+ form, while the lower part (2.5 cm height from the starting line) is in the Na^+ form. Gln and Asn can be separated sharply from the other amino acids, such as Asp, Glu, Gly, Thr, Ser developed on a chromatosheet in such a "tandem" ionic form with a lithium citrate/formate buffer. If the whole layer is in the Li^+ form and the same developing buffer is applied, all amino acids in the mixture can be separated, except Gln and Glu which form a common spot. On the other hand, if the whole layer is in its original Na^+ form, only Gln and Glu can be separated under similar experimental conditions, while the other amino acids produce a diffuse, common spot. These observations led to the application of the "tandem" ionic form of resin-coated chromatosheets for the one-dimensional separation of amino acid mixture.

H. Separation and Determination of Methionine Sulfone in Hydrolysates of Samples Treated with Performic Acid

The determination of the exact amount of methionine is a critical task in both protein chemistry and plant breeding. This amino acid is unstable during acid hydrolysis and is decomposing to a certain extent (forming sulfoxide and sulfone). Methionine can be quantitatively converted to the stable sulfone form ($MeSO_2$) by performic acid treatment.

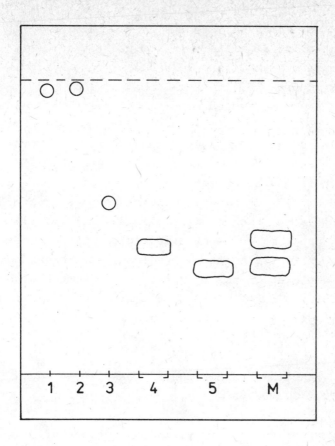

Fig. 7. Racemization test for peptide synthesis. Chromatography
on strongly acidic resin-coated chromatosheet (Fixion
50X8 in Na^+-form). Developing buffer: pH = 4.25,
Na^+ = 0.8 mol/l, citrate = 0.07 mol/l. Ingredients per
liter: citric acid monohydrate = 14.1 g, sodium
hydroxide = 8.0 g, sodium chloride = 35.0 g, 37% hydrochloric acid (sp.gr. 1.19) = 8.4 ml. Equilibration by
continuous development for 25 hours at room temperature
with a 10-fold diluted developing buffer. Chromatography at room temperature. Solvent front height:
approx. 10-12 cm. Sampes applied as 1 cm band (10 μl
each from a solution containing 10 mg per ml peptide
material).

1 = GLY,
2 = ALA,
3 = LEU,
4 = L-L GLY.ALA.LEU,
5 = D-L-Gly.ALA.LEU
M = 1:1 mixture of 4 + 5

When the number of samples to be analyzed daily is high, the operation of an amino acid analyzer is not very economic, and its capacity is limited. In such a case a simple TLC method has great importance.

When several samples have to be analyzed simultaneously the capacity of the amino acid analyzer may be the limiting step of the program. In such cases a simple TLC procedure can overcome this difficulty. Váradi and Pongor (1979) developed a simple, one-dimensional method for the detection and determination of $MeSO_2$ in the hydrolysates of samples treated with performic acid. This technique is also suitable for large-scale screening programs in plant breeding for the detection of high methionine mutants. A strongly acidic resin-coated chromatosheet is employed in the Li^+ form and developed with a lithium citrate buffer, pH = 2.50 ± 0.55 (Li^+ = 0.6 mol/l). $MeSO_2$ forms a single spot between Glu and Thr/Ser.

1. THIN-LAYER ION-EXCHANGE CHROMATOGRAPHY
   OF ACIDIC COMPOUNDS

The determination of cystine cannot be carried out directly from acidic hydrolysates since the former amino acid completely decomposes under acidic hydrolytic conditions. Cystine, however, can be transformed into the acid-stable cysteic acid ($CYSO_3H$) by oxidation with performic acid. The latter strongly acidic amino acid derivative can be readily separated from all the other amino acids by chromatography on strongly <u>basic</u> resin-coated chromatosheets (Fixion 2X8 in acetate form). Cysteic acid is strongly bound to the basic resin present in the thin layer and it has the lowest R_f value among all the other amino acids (Fig. 8). Therefore, relatively small amounts of $CYSO_3H$ can be detected by overloading the chromatosheet with a large excess of other amino acids (Ferenczi and Dévényi 1971, Váradi and Pongor 1979). Due to the rapidity and simplicity of this technique, it is suitable for the screening of large number of samples in sequence analysis or in plant breeding programs.

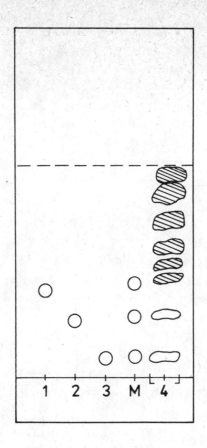

Fig. 8. Separation of cysteic acid in hydrolysates of performic
acid treated peptides or proteins on strongly basic
resin-coated chromatosheet (Fixion 2x8 in acetate-form).
Equilibration: continuous development for 24 hours at
room temperature with 0.01 mol/l acetic acid. Developing buffer: pH = 3.8. Ingredients per liter: pyridine
10 ml, acetic acid 100 ml. Solvent front height = 15 cm.
Samples containing traces of $CYSO_3H$ should be applied
as 1 cm bands.
1 = GLU, 2 = ASP, 3 = $CYSO_3H$, M = Mixture of 1-3,
4 = acid hydrolysate of a performic acid treated pea-
sample.
Comment: Clear separation can also be achieved with a
sodium acetate buffer pH = 3.8, Na^+ = 0.02 mol/l,
acetate = 0.2 mol/l.
Preparation of the buffer: (a) 27.2 g/l sodium acetate,
(b) 12 ml acetic acid/l. Buffer = 10 ml (a) + 90 ml (b).

Fábián employed this method for the investigation of the
N-terminal group of pig pancreatic amylase. Pyroglutamic acid
was shown to be the N-terminal of the enzyme.

Himoe and Rinne (1978) applied this chromatographic method for the separation of fumarate, citrate, malate, aspartate and glutamate from crude extracts, eliminating the need of pretreatment of the extracts with ion-exchange columns or the use of a two-dimensional electrochromatographic technique.

2. ONE-DIMENSIONAL SEPARATION OF BIOGENIC AMINES ON RESIN-COATED CHROMATOPLATES

Biogenic amines are frequently analyzed in foods and feeds where they are present as decomposition products of amino acids. They are therefore important as subject of quality control in the food and feed industry, as the markers of unwanted side-reactions during processing and storing.

The amines have also received much attention on a quite different field: in carcinogenesis studies. Several authors have described the effect of carcinogens on hepatic polyamines.

Seiler (1977) reviewed the methods for the detection and determination of polyamines. Mainly TLC, GLC or column chromatographic techniques are described.

Lepri et al. (1978) studied the separation of amines on home-made mixed layers. The separation of these compounds was found to be due to partitioning between the two phases and not to an ion-exchange mechanism.

More recently two procedures were published employing precoated resin layers (Fixion 50X8 in Na^+ form).

Procedure A. (Pongor et al. 1980):

a) Developing buffer: sodium citrate pH 6.0, Na^+ = 2.5 mol/l. Ingredients: sodium citrate dihydrate 19.6 g/l, sodium chloride 134.4 g/l The pH is adjusted to 6.0 with 37% HCl.

b) Development at room temperature.

c) Detection with ninhydrin-cadmium spray reagent.

d) Quantitative estimation by video-densitometry in the range of 1-7 µg/spot.

e) Elution order of the amines in increasing order of R_f values: tyramine, agmatine, histamine, cadaverine, putrescine, diaminopropane.

560

Procedure B (Bardócz and Karsai 1981):

a) Developing buffer: sodium phosphate pH 7.5. Ingredients: 200 mmol/l potassium hydrogenphosphate, 2 mol/l sodium chloride. The pH is adjusted to 7.5 with NaOH.

b), c) and d) are the same as detailed in Procedure A.,

e) Elution order of the amines under the above mentioned experimental conditions: spermine, agmatine, spermidine, putrescine, arginine, ornithine, methylamine.

Bardócz et al. (1982) measured and compared the polyamine content of the liver and kidney in newborn female rats injected once with dimethylnitrosamine. The authors found a close relationship between the concentration of polyamines and histological changes during tumor development.

3. APPLICATION OF THIN-LAYER ION-EXCHANGE CHROMATOGRAPHY IN CLINICAL BIOCHEMSITRY

As far as clinical application is concerned, a simple one-dimensional separation method for amino acids may have an important role in

a) early detection of amino acid disorders,

b) control of therapy (e.g. protein restriction), and

c) enzyme assays (amino acid metabolism).

Thin-layer ion-exchange chromatography is well suited for an early detection of hereditary metabolic disorders of newborn infants. These metabolic disorders are characterized by an accumulation of certain amino acids in blood or urine, because of genetically defined enzymatic damage. Thin-layer ion-exchange chromatography is a simple, easy method, because:

a) Acids, salts, etc. can be present in the sample without any interference of the separation of amino acids;

b) all separations can be carried out by one-dimensional runs, i.e. a large number of the samples can be handled simultaneously (Pongor et al. 1978b);

c) combining the method with videodensitometry, the amount (or concentration) of the amino acids can be measured from very small quantities of tissue materials (biopsy), cells or body fluids.

The most common disorder of amino acid metabolism is phenylketonuria (PKU). There are other diseases that cause an accumulation of lysine, histidine, tyrosine, valine, isoleucine and leucine. Cystinurea is characterized by the presence of a non-natural amino acid: homocystine (in addition to the accumulation of cystine). If the diseases are recognized in the first days of the infants' life, they can be cured by a suitable diet over the years. In the opposite case, the patients become mentally retarded. The screening of PKU was made compulsory in many countries.

Thin-layer ion-exchange chromatography has the major advantage of offering a simple technique to characterize practically any disorders of amino acid metabolism from a single sample with a one-dimensional run. Considering the separation of amino acids in Fig. 3, it is evident that any of the possible disorders listed in Table II (except of cystinurea) can be detected from a one-dimensional chromatogram developed at pH 5.1. It is possible to detect phenylketonurea, lysineaemia, histidinaemia, tyrosinaemia and the Maple Syrup Urine Disease (a disorder of the branched-chain amino acids) from the elevated level of Phe, Lys, His, Tyr, Val/Ile/Leu, respectively.

Some results concerning the early detection of amino acid disorders by thin-layer ion-exchange chromatography on resin-coated chromatosheets are listed in Table II.

4. APPLICATION OF THIN-LAYER ION-EXCHANGE CHROMATOGRAPHY
   FOR THE MEASUREMENT OF THE ACTIVITY OF AMINO ACID
   TRANSFORMING ENZYMES

The measurement of the enzyme activities in small tissue samples (received by biopsy) has a great importance in clinical diagnostics. Recently Karsai et al. (1978, 1979) and Karsai and Elődi (1979, 1980) have elaborated micro methods for the measurement of the activity of enzymes related to amino acid metabolism. The basic principle of the method is the following:

a) A few mg of a tissue specimen is taken by biopsy.

b) A homogenate (or tissue extract) is made from the sample with a final volume of about 50-100 µl.

c) At appropriate time intervals 5-10 µl aliquots are taken and spotted directly to a resin-coated chromatosheet.

d) Selecting the appropriate buffer the amino acid(s) to be estimated are separated by one-dimensional chromatography.

e) The change in the concentration of the selected amino acid (participating in the enzyme reaction) is estimated by video-densitometry or with an equivalent method.

Changes in the amino acid content as low as 1 to 2 nmol can be reliably detected. Enzymes listed in Table III can be estimated with reasonable accuracy from small amounts of tissue samples employing this micro method.

5. APPLICATION OF THIN-LAYER ION-EXCHANGE CHROMATOGRAPHY
   FOR THE DETERMINATION OF ESSENTIAL AMINO ACIDS
   IN PLANT PROTEINS

The concentration of the nutritionally limiting amino acids (first of all lysine, methionine, tryptophane and cystine) is a direct index of the nutritional quality. It is used as a coarse ranking indicator in plant breeding programs. Several methods were elaborated for the determination of these amino acids, but only a few methods can be employed on large scale. Usually the dye-binding method (DBC) is applied which is an un-specific, comparative method for the determination of the total amount of basic amino acids (the sum of lysine, histidine and arginine), and there is no accepted method in the literature for the large-scale estimation of other limiting amino acids (i.e. methionine, tryptophane and cystine).

Thin-layer ion-exchange chromatography on strongly acidic, resin-coated chromatosheets makes it possible to separate each limiting amino acid with a one-dimensional (single) run. A few ul of the acidic hydrolysate of the plant seed is chromato-graphed according to one simple method described in Section II, the chromatograms are visualized by ninhydrin and subsequently

Table I. One-dimensional separation methods on strongly acidic resin-coated chromatosheets or plates(*)

pH	Ionic form	Buffer	Compounds	Reference
2.3	H^+	acetic acid/HCl	alkaloids	Lepri et al. 1976b
2.4	Li^+	lithium-citrate	$MeSO_2$ from other amino acids	Váradi and Pongor 1979
2.8	Na^+/Li^+ (tandem)	lithium-citrate/formate	Asn and Gln form other amino acids	Váradi 1975
3.3	Na^+	sodium citrate	16 amino acids (except Ser/Thr)	Dévényi et al. 1971b, Dévényi 1982
4.2	Na^+	sodium citrate	aromatic + basic containing Orn	Issaq and Dévényi 1981
4.4	Na^+	sodium citrate	D-Penicillamine	Pongor et al. 1978c
5.1	Na^+	sodium citrate	basic + aromatic	Dévényi 1976

(Table I. cont.)

pH	Ionic form	Buffer	Compounds	Reference
5.28	Na$^+$	sodium citrate	hydroxylysine	Buzás and Boross 1980
6.0	Na$^+$	sodium citrate	tryptophan	Dévényi et al. 1971
6.0	Na$^+$	sodium citrate	biogenic amines	Pongor et al. 1980
6.5	Na$^+$	sodium phosphate	Nebramycins	Kádár-Pauncz 1978
7.0	Na$^+$	sodium acetate	sulfonamides	Kádár-Pauncz 1981
7.5	Na$^+$	potassium phosphate	polyamines	Bardócz and Karsai 1981
8.5	Na$^+$	sodium acetate	basic antibiotics	Kádár-Pauncz 1972

(*) Table I does not include the solvents used for the one-dimensional separation of different nucleic acid constituents discussed in Section IV.

Table II. Detection of amino acid disorders by thin-layer
ion-exchange chromatography on resin-coated chromatosheets

Result of screening	Reference
PKU	Kovács 1973
Hyperlysineaemia	Szabó et al. 1975
PKU, cystinurea, histidinaemia, tyrosinaemia	Vargáné-Szabó 1976
Heterozygosity in aminoacidopathia	Tasnádi et al. 1977
Amino acid composition of amniotic fluid in early pregnancy	Juhász et al. 1977
Mild variant of Maple Syrup Urine Disease	Kovács and Kiss 1978, Pongor et al. 1978b
Maple Syrupe Urine Disease	Kovács 1979
Hyperalaninaemia	Kovács 1981
Dibasic amino aciduria (carbamoyl-phosphate synthetase-defect	Szentpéter et al. 1982
Screening of 372.000 newborns: PKU = 1 : 6000 Hyperphenylalaninaemia 1 : 35000 two cases of tyrosinaemia	Zukerman 1982[*]

(*) Zukerman, G.L. (1982) (Mogilev, Bielorussia, USSR): Mass
and Selective Identification of Hereditary Metabolic
Disorders in Bielorussia (USSR) by Chromatography on Fixion
50X8 Chromatosheets. Reported at the Fifth Symposium of the
Socialist Countries on Hereditary Metabolic Disorders,
Moscow, May 18-20, 1982.

Table III. Micro-assays of amino acid transforming enzymes
           applying thin-layer ion-exchange chromatography

Enzyme	Change in amino acid content	Reference
Glutamate dehydro-genase (EC 1.4.1.3)	Increase of Glu	Karsai and Elődi 1979
Ornithine-2-oxoacid amino-transferase (EC 2.6.1.3)	Increase of Glu	ibid
Histidine ammonia-lyase (histidase, EC 4.3.1.3)	Decrease of His	ibid
Carbamoyl-phosphate-synthetase (EC 2.7.2.5)	Increase of rulline	Karsai et al. 1979
Arginino-succinate lyase (EC 4.3.2.1)	Decrease of ar-ginine-succinate	ibid
Arginase (EC 3.5.3.1)	Increase of Orn	ibid
Ornithine carbamoyl-transferase (EC 2.1.3.1)	Increase of cit-rulline	ibid

used for quantitative estimation with video-densitometry. A
detailed experimental schedule has been elaborated at the Joint
Laboratory of the International Atomic Energy Agency (IAEE) and
Food and Agricultural Organization of the United Nations (FAO)
at Seibersdorf, Austria. The daily output of lysine determination from different plant varieties is usually 600-800 analysis
in this Laboratory. 16,000 wheat lines were analyzed in 1977
and a mutant was selected containing about 30% more lysine than
the control variety (Sigurbjörnsson et al. 1979).

The basic principle of the large scale screening of amino acids is the following:

The hydrolysates of the different plant seeds are chromatographed under strictly controlled experimental conditions. Each chromatogram contains one standard sample (which has been carefully analyzed for its limiting amino acid content, e.g. for lysine) and 5 to 6 unknown samples. The spot density integrals (for e.g. lysine) and that of the total amount of amino acids are measured simultaneously. The spot density integral of lysine and that of the total amino acid content are linearly related to the sample size. However, their ratio (L%) is independent of the sample size within a broad range (Fig. 9). Consequently, accurate weighing and sample application are not necessary, provided that the amount of sample per spot is within this range (Dévényi 1976a, 1976b, Pongor et al. 1976, Dévényi et al. 1978).

Thus, sample preparation is made simple and productive, which is the basic reason why the method can be economically applied in large scale screening programs. An International Laboratory Comparison was organized by the IAEA in 1978. Eight different methods for the determination of lysine were compared: video-densitometry after separation on ion-exchange thin-layer chromatography, infrared analysis, amino acid analyzer technique, DBC method, fluorescent method (after dansylation), enzyme hydrolysis, electrophoretic procedure and a spot test combined with chromatography. The reproducibility of the methods in each laboratory was assessed from the variation of the analytical results from replicates of a check genotype of each species. Video-densitometry (combined with thin-layer ion-exchange chromatography) appeared to be the most satisfactory method (Georgi et al. 1979).

Thin-layer ion-exchange chromatography combined with video-densitometry was applied by several laboratories for the large scale screening of lysine contents in different plant varieties (wheat: Belea and Sági 1978, Dévényi et al. 1978, Rao et al. 1979, Sági et al. 1980, Brunori et al. 1980; Millet: Rabson et al. 1978; Barley: Winkler and Schön 1978; Maize: Szirtes et al. 1977; Grape: Juhász and Polyák 1976).

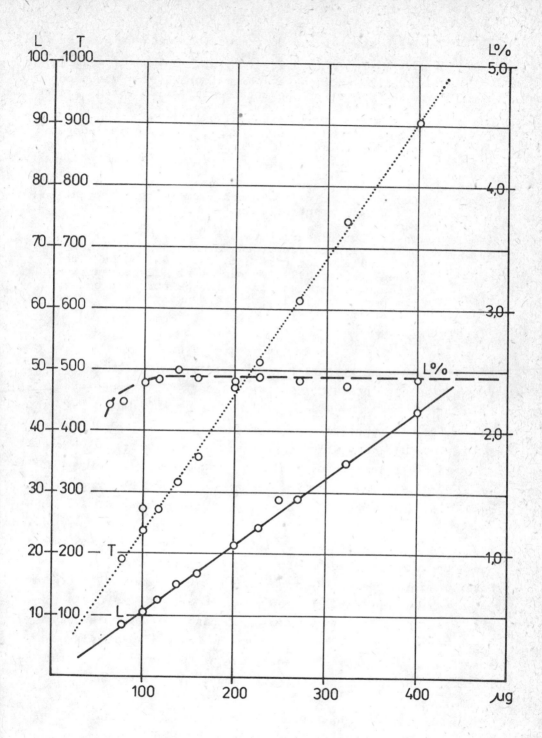

Fig. 9

Fig. 9. Determination of lysine content of wheat-samples by
thin layer ion-exchange chromatography and video-
densitometry (Pongor 1982). L = measured integral value
for lysine spots on the chromatograms, T = measured
integral value for the total amino acid content present
on the chromatograms. L% is the calculated ratio
between L and T. The integral values for L and T are
linearly related to sample size, while L% is independent from the latter in a relatively broad range.

Comments:

a) Video-densitometry can be substituted by any other
densitometric method which permits the integration of
the total amount of amino acids present in the
sample.
b) The measured density values can be converted
directly into percentual values (W/W) with the aid
of a calibration constant, which should be determined individually for each type of substance.
Through multiplication by the calibration constant,
the percentual value of the density can be converted
to percentage W/W.

CHROMATOGRAPHIC BEHAVIOUR OF NUCLEIC ACID CONSTITUENTS ON
STRONGLY ACIDIC, RESIN-COATED CHROMATOSHEETS

Different types of ion-exchangers are widely used for the
chromatographic separation of nucleic acid constituents. The
ion-exchange column-chromatographic techniques were introduced
by Cohn (1975); the topic was recently reviewed by Cohn (1975),
while Lin (1974) published a survey of the automated analyzer
technique.

Since the 1960s thin-layer chromatography has become an
indispensable micro method for the separation of nucleic acid
bases, nucleosides, nucleotides and related compounds. Several
reviews have been published on the methodology of TLC and the
main results in this field are concluding that anion-exchange
chromatography, first of all employing PEI-cellulose plates or
sheets, has the greatest importance in the separation of these
compounds (Randerath 1967, Pataki 1968, Böhme and Schultz
1974).

Much less work has been done on thin-layer cation-exchange
chromatography (Randerath and Randerath 1963, Lepri et al.
1972). Since the early seventies systematic studies have been

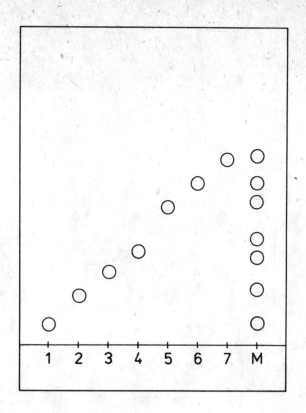

Fig. 10. Separation of RNA-bases on strongly acidic resin-
coated chromatosheet (Fixion 50X8 in H$^+$-form). The
chromatosheets in Na$^+$-form are converted to H$^+$ by
continuous development at room temperature for 16
hours with 1.0 mol/l HCl. Developing solution: HCl,
1.0 mol/l., at room temperature (approx. 90 minutes).

1 = adenine, 2 = guanine, 3 = cytosine, 4 = hypoxanthine, 5 = xanthine, 6 = thymine, 7 = uracil, M =
= mixture of 1-7. Tracing showing the relative spot
positions was drawn on the basis of the original work
of Tomasz (1973).

performed in the laboratory of Tomasz on the chromatographic
behaviour of nucleic acid bases, nucleosides, nucleotides and
some related compounds on strongly acidic (cationic) resin-
coated chromatosheets (Tomasz 1973, 1974a,b, 1975, 1976, 1979a,
b, 1980a,b, Tomasz and Farkas 1975, Tomasz and Simonovics 1975).
These investigations led to the development of several interesting separation methods and a possible explanation of the
cation ion-exchange mechanism of these compounds could be
deducted from the studies (Tomasz 1980a).

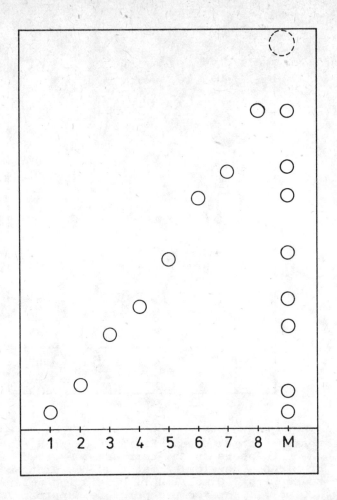

Fig. 11. Separation of DNA-bases on strongly acidic resin-
coated chromatosheets in H^+-form. The chromatosheets
in Na^+-form are converted to the H^+-form by continuous
development with 2.8 mol/l hydrochloric acid, at room
temperature for 16 hours. Developing solution: 2.8
ml/l hydrochloric acid. Chromatography at room temperature for 5 hours.
1 = N^6-methyl adenine, 2 = adenine, 3 = guanine,
4 = 5-methyl cytosine, 5 = cytosine, 6 = 5-hydroxy-
methyl cytosine, 7 = thymine, 8 = uracil, M = mixture
of 1-8.
Comment: The separation of the bases is highly
pH-dependent.
The tracing was drawn on the basis of the original
work of Tomasz (1975).

The separation and determination of nucleic acid constituents (except a very few cases) are usually carried out in

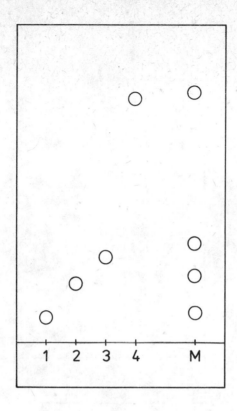

Fig. 12. Separation of nucleotides on strongly acidic resin-
coated chromatosheets (Fixion 50X8 in H$^+$-form). The
sheets in Na$^+$-form are converted to H$^+$-form by continuous developing at room temperature, for 16 hours
with 1.0 mol/l HCl. The excess of HCl should be
eliminated by continuous development with deionized
water.
Chromatography is carried out by developing with
deionized water, at room temperature.
1 = adenosine monophosphate, 2 = cytidine monophosphate, 3 = guanosine monophosphate, 4 = uridine
monophosphate, M = mixture of 1-4.
The tracing showing the relative spot positions was
drawn according to Tomasz (1973).

the H$^+$ form of the strongly acidic resin-coated chromatosheet
(Fixion 50X8) instead of the Na$^+$ or Li$^+$ form, the latter cur-
rently employed for the separation of amino acids. Such
chromatosheets can be easily prepared from the sheets available
in the Na$^+$ form by continuous development with diluted hydrochloric acid.

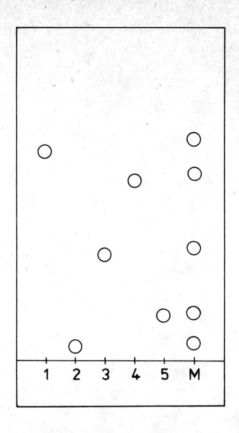

Fig. 13. Separation of adenosine-derivatives on strongly
acidic resin-coated chromatosheets (Fixion 50X8 in Na^+-
form). Equilibration: continuous development at
room temperature for 16 hours with a solution of
sodium bicarbonate 0.01 mol/l. Developing solution:
sodium bicarbonate, 0.1 mol/l.
1 = adenosine, 2 = 1-methyl-adenosine, 3 = 6.2-
dimethyl adenosine, 4 = 6-methyl adenosine,
5 = 6-isopentenyl-adenosine, M = mixture of 1-5.
The tracing showing the relative spot positions
was drawn on the basis of the original work of
Tomasz (1979b).

The separation of the RNA-bases (adenine, guanine,
cytosine, hypoxanthine, thymine and uracil) (Fig. 10) can be
performed with a one-dimensional (single) run using 1.0 mol/l
hydrochloric acid (Tomasz 1973), while a higher concentration
of hydrochloric acid (2.8 mol/l) is convenient for the one-
dimensional separation of the DNA-bases (Fig. 11) according to
Tomasz (1975). The separation of the DNA-bases is strongly pH

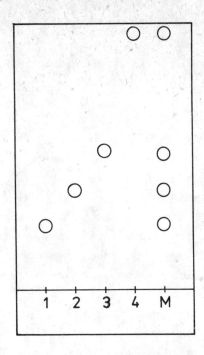

Fig. 14. Separation of ribonucleoside 3',5'-cyclic phosphates
on strongly acidic resin-coated chromatosheets Fixion
50X8 in H^+-form). The chromatosheets in Na^+-form are
converted into H^+ form by continuous development with
2.0 mol/l hydrochloric acid at room temperature, for
16 hours. Developing solution: oxalic acid, 0.05
mol/l. Solvent front height: 10 cm (approximatively
90 minutes).
1 = adenosine-3',5'-cyclic phosphate, 2 = cytidine-
3',5'-cyclic phosphate, 3 = guanosine-3',5'-cyclic
phoshate, 4 = uridinine-3',5'-cyclic-phosphate,
M = mixture of 1-4.
The tracing was drawn on the basis of the original
work of Tomasz (1979a).

dependent. Even a small change in the HCl concentration has a
great influence on the separation of N^6-methyladenine from
adenine. This simple one-dimensional method was employed by
Kirnos et al. (1977) for the detection of 2-amino-adenine as an
adenine substituting base in S-2L cyanophage DNA.

Distilled water can be used as the eluting solvent for the
one-dimensional cation-exchange TLC of nucleotides on strongly
acidic resin-coated chromatosheets in H^+ form (Fig. 12).
However, in this case, following the change of the ionic form

from Na$^+$ to H$^+$ by continuous development with diluted hydrochloric acid, the excess of acid must be removed by a second
prewash using distilled water as eluent (Tomasz 1973). This
method combined with ligand-exchange chromatography and gel
filtration was applied for the purification and identification
of the free nucleotides from Euglena gracilis 2 (Krauss and
Reinbothe 1977).

The one-dimensional cation-exchange chromatographic separation of some adenosine derivatives (Fig. 13) is one of the very
few examples in which a resin-coated chromatosheet in Na$^+$ form
can be employed (Tomasz 1979b). The eluting solution was 0.1
mol/l sodium bicarbonate. Prior to use, the chromatosheets have
to be equilibrated by continuous development with 10-fold
diluted eluting solution for 16 hours at room temperature.

The one-dimensional separation of RNA-cyclic phosphates
(Fig. 14) can be carried out on strongly acidic resin-coated
chromatosheets in H$^+$ form. The developing solvent is 0.05 mol/l
oxalic acid (Tomasz 1979b).

Cation-exchange thin-layer chromatography of nucleic acid
constituents was adapted also for enzyme assays. Tomasz and
Farkas (1975) described a micro method for the determination of
adenyl cyclase activity. Dusha and Dénes (1977) applied cation-
exchange TLC on resin-coated chromatosheets for the determination of the activity of enzymes catalyzing ATP-pyrophosphate
exchange reactions. This method offers an easy way for the
simultaneous determination of large series of small samples.

ONE-DIMENSIONAL SEPARATION OF ALKALOIDS BY THIN-LAYER
ION-EXCHANGE CHROMATOGRAPHY

Lepri et al. (1976a,b) investigated in detail the chromatographic behaviour of 48 different alkaloids by thin-layer ion-
exchange chromatography on home-made anionic and cationic
chromatoplates.

The authors listed the R_f values of the alkaloids on
AG1-X4 chromatoplates (in acetate form) prepared from a mixture
of the resin (3 g, rinsed with water, methanol and dried at

Table IV.   One-dimensional separation of alkaloids on AG1-X4
chromatoplates eluted with 0.5 mol/l ammonium
acetate[*]

R_f	Alkaloids
0.00	Nacreine Ergocristine Ergotamine
0.02-0.04	Papaverine Jbogaine Berberine hydrochloride Reserpine
0.07-0.08	Boldine Ergonovine
0.22	Narcotine
0.36-0.38	Hydrastine Aminophylline Theophylline
0.42-0.44	Colchicine Yohimbine hydrochloride
0.50-0.54	Quinine Quinidine sulphate
0.66-0.68	Brucine Theobromine Caffeine
0.71	Cinchonine hydrochlorid Cinchonidine
0.77-0.82	Strychnine Ajmaline Lobeline hydrochloride

(Table IV, cont.)

R_f	Alkaloids
0.85	Tubocurarine
0.90-0.91	Cocaine
	Atropine
	Hyoscyamine
	Eucatropine hydrochloride
	Emetine hydrochloride
	Ethylmorphine
0.93-0.94	Eserine sulphate
	Hometropine
	Scopolamine hydrochloride
	Arecoline hydrochloride
0.96-0.98	Hyoscyne
	Scopoline
	Sparteine sulphate
	Prostigmine

(*) Based on data from Lepri et al. 1976a

room temperature prior to use) and of microcrystalline cellulose (9 g) in 50 ml water. Detection of the alkaloids was performed by fluorescence, except cocaine, which was detected by the Drangendorff reagent.

The amount of alkaloids varied from 0.25 µg to 10 µg per spot which is in general lower than that detected by other chromatographic methods. Based on the data of the authors, the R_f values of the investigated alkaloids can be divided into 15 major groups (Table IV).

A very sharp one-dimensional separation can be obtained in the case of eight alkaloids (which do not migrate or have very low R_f values in the case of development with 0.5 mol/l ammonium acetate) using 1 mol/l acetic acid as the eluting solution. The elution sequence of these alkaloids is (in order of increasing

Table V. R_f values of alkaloids on cationic thin-layers in H^+-form

Alkaloid	R_f values on		
	Rexyn 102[*]	Dowex 50X4[**]	Alginic acid[*]
Spermine	0.06	–	0.01
Spermidine	0.11	–	0.01
Quinine	0.02	0.00	0.02
Quinidine sulphate	0.02	0.00	0.02
Cinchonine hydro-chloride	0.03	0.00	0.03
Cinchonidine	0.03	0.00	0.03
Tubocurarine	0.01	0.00	0.04
Ergocristine	0.00	0.01	0.04
Berberine hydro-chloride	0.00	0.00	0.05
Emetine hydrochloride	0.01	0.00	0.05
Reserpine	0.00	0.00	0.06
Ergotamine	0.01	0.00	0.06
Sparteine sulphate	0.18	0.04	0.08
Narceine	0.03	0.01	0.12
Ergonovine	0.02	0.05	0.12
Boldine	0.02	0.00	0.15
Yohimbine hydrochloride	0.03	0.02	0.15
Strychnine	0.03	0.01	0.17
Brucine	0.03	0.01	0.18
Jbogaine	0.01	0.00	0.19
Papaverine	0.02	0.00	0.21
Hydrastine	0.02	0.01	0.22
Narcotine	0.05	0.02	0.23
Ajmaline	0.08	0.01	0.27
Morphine	0.23	0.13	0.27
Lobeline hydrochloride	0.02	0.01	0.28
Scopoline	0.42	0.32	0.28
Tropine	0.42	0.28	0.29
Ethylmorphine	0.13	0.04	0.29

(Table V. cont.)

Alkaloid	R_f values on		
	Rexyn 102 [(*)]	Dowex 50X4 [(**)]	Alginic acid [(*)]
Cocaine	0.07	0.05	0.30
Scopolamine hydro-chloride	0.21	0.18	0.31
Eserine sulphate	0.12	0.01	0.32
Arecoline hydro-chloride	0.33	0.17	0.33
Atropine	0.14	0.07	0.33
Hyoscyamine	0.14	0.08	0.33
Homatropine	0.18	0.10	0.33
Eucatropine hydro-chloride	0.07	0.03	0.34
Hyoscine	0.07	0.03	0.34
Prostigmine	0.17	0.13	0.39
Theophyline	0.35	0.29	0.69
Aminophyline	0.35	0.28	0.69
Theobromine	0.36	0.24	0.76
Caffeine	0.28	0.20	0.81
Colchicine	0.08	0.01	0.91

[(*)] Eluted with 1 mol/l acetic acid + hydrochloric acid at pH = 2.35

[(**)] Eluted with 1 mol/l hydrochloric acid

[(***)] Based on the data of Lepri et al. 1976b.

R_f values): narcein, berberine, ergonovine, reserpine, ergocristine, jbogein, quinine, cocaine.

Lepri et al. (1976b) studied the chromatographic behaviour of alkaloids also employing cation exchangers. Alginic acid, Rexyn 102, Dowex 50X4 and CM-cellulose were used for the preparation of the home-made ion-exchange chromatoplates. The

CM-cellulose layers were prepared in the following way: 4 g of the CM-cellulose was mixed with 50 ml water.

The R_f values of alkaloids on three different cationic chromatoplates eluted with acidic solvents are summarized in Table V, based on the data of Lepri et al. (1976b). In the presence of a large amount of alcohol (30 to 50%) the retardation of alkaloids is markedly decreasing. By higher alcohol concentration (buffer : alcohol = 1:1 or 1:4) partition and/or adsorption phenomena are prudent instead of ion exchange mechanism.

APPLICATION OF THIN-LAYER ION-EXCHANGE CHROMATOGRAPHY FOR THE SEPARATION OF ANTIBIOTICS, SULFONAMIDES, AND SOME OTHER CHEMO-THERAPEUTIC COMPOUNDS

Various methods have been reported for the separation and determination of water-soluble basic antibiotics (Vondracek 1975). Only few methods were based on ion-exchange chromatography (Inouye and Ogawa 1964). Most authors applied TLC and/or electrophoresis on silica gel, employing multiple developments or two dimensional combinations.

Kádár-Pauncz (1972) described a simple method for the one-dimensional separation of deoxystreptamine-containing (basic, water soluble) antibiotics as well as other related compounds of different structure.

The R_f values of basic, water-soluble antibiotics and related compounds are summarized (Kádár-Pauncz 1972) in Table VI.

TLC on resin-coated chromatosheets has been employed also for the separation and identification of the nebramycin components (Kádár-Pauncz 1978). Streptomyces tenabrarius produces a variety of nebramycins. The complex can be separated by paper chromatography with a 40 hours of development (Higgins and Kastner 1967) but using Fixion 50X8 in Na$^+$ form, the separation was completed in 7 hours (Kádár-Pauncz 1978).

A great number of sulfonamides and related substances can be separated on strongly acidic resin-coated chromatosheets

Table VI. R_f values of basic, water-soluble antibiotics and related compounds on strongly acidic resin-coated chromatosheets at pH = 8.5 [*]

Compound	R_f
Polymyxin B	0.00
Bacitracin	0.01
Gentamycin	0.05
Capreomycin	0.05
Lincomycin	0.14
Neomycin	0.15
Viomycin	0.25
Neamine	0.28
Paromomycin	0.32
Streptidine	0.37
Kanamycin	0.39
Dihydrostreptomycin	0.44
Hygromycin B	0.45
Streptomycin	0.49
Paromamine	0.52
Deoxystreptamine	0.59
Streptamine	0.65
Mannosidohydroxystreptomycin	0.66
Cycloserine	0.76
Glucosamine [**]	0.83

[*] Based on the data of Kádár-Pauncz 1972.

[**] The separation of glucosamine from galactosamine and some amino acids can also be performed by the method of Hrabák (1973) employing a sodium citrate buffer, at pH = 5.28.

(Fixion 50X8 in Na^+ form) at pH 7.0 (Kádár-Pauncz 1981). The composition of the eluting buffer is 1.6 mol/l sodium acetate + 1 mol/l sodium chloride + 20% ethanol, the pH is adjusted to 7.0. The R_f values of some sulfonamides and related substances are summarized (Kádár-Pauncz 1981) in Table VII.

Table VII. R_f values of sulfonamides and related compounds [*]

Compounds	R_f
Diaveridin	0.07
Ormetoprim	0.08
Sulfquinoxaline	0.12
Sulfpiride + Trimethoprim	0.16
Sulfadimethoxine	0.18
Sulfamethylthiazole	0.20
Sulfamethoxyparidazine + Sulfathiazole	0.23
2-Methoxy-5-sulfanilamine-methyl benzoate	0.29
Sulfacytine	0.30
Sulfadimidine	0.31
Sulfamethoxazole	0.38
Sulfamethoxypyrazine	0.39
Acetylsulfaguanidine + Sulfanilamide	0.40
Sulfaguanidine	0.47
Sulfacetamide	0.74
Sulfathiourea	0.79
Actylsulfathiourea	0.80
Sulfanilic acid	0.90

[*] Based on the data of Kádár-Pauncz 1981.

CONCLUDING REMARKS

The development of biochemical analysis obviously tends towards automated computer-controlled autoanalysers. This trend has already been shown by the increasing number of commercially available, although expensive, sophisticated microprocessor controlled analyzers. Yet simple and versatile techniques such as TLC are not discarded. It is the authors' belief that these productive and very informative methods will not lose their practical importance in the foreseeable future.

REFERENCES

Bardócz, S. and Karsai, T.: J. Chromatogr. 223, 198-201 (1981)

Bardócz, S., Várvölgyi, Cs., Rády, P., Nagy, A. and Karsai, T.: Anticancer Research 2, 309-314 (1982)

Belea, A. and Sági, F.: Cereal Res. Comm. 6, 15-19 (1978)

Biszku, E., Sajgó, M., Solti, M. and Szabolcsi, G.: Eur. J. Biochem. 38, 283-292 (1973)

Böhme, E. and Schultz, G.: in Methods in Enzymology (Collowick, S.P. and Kaplan, N.O., Eds.) Vol. 38., Academic Press, New York (1974), pp. 27-38

Brunori, A., Axmann, H., Figueroa, A. and Micke, A.: Z. Pflanzenzüchtg. 84, 201-218 (1980)

Buzás, Zs., Polyák, B. and Boross, L.: Acta Biochim. Biophys. Acad. Sci. Hung. 15, 173-176 (1980)

Cardari, V., Lederer, M. and Ossicini, L.: J. Chromatogr. 101, 411 (1974

Clarke, S.D., Romsos, D.R., Tsai, A.C., Belo, P.S., Bergen, W.G. and Leveille, G.A.: Nutrition 106, 94-102 (1976)

Cohn, W.E.: in The Nucleic Acids (Chargaff, E. and Davidson, J.N., Eds.) Academic Press, New York (1955) p. 211

Cohn, W.E.: in Chromatography (Heffmann, E., Ed.), Van Nostrand Reinhold Company, New York (1975) pp. 714-743

Cosrba, I., Buzás, Zs., Polyák, B. and Boross, L.: J. Chromatogr., 172 287-293 (1979)

Dávid, A. and Takácsy, T.E.: J. Pharm Pharmacol. 27, 774-775 (1975)

Dévényi, T.: Acta Biochim. Biophys. Acad. Sci. Hung. 5, 435-440 (1970a)

Dévényi, T.: Acta Biochim. Biophys. Acad. Sci. Hung. 5, 441-445 (1970b)

Dévényi, T.: Acta Biochim. Biophys. Acad. Sci. Hung. 11, 1-10 (1976a)

Dévényi, T.: Hung. Sci. Instr. 36, 1-6 (1976b)

Dévényi, T.: in Advances in Thin Layer Chromatography (J.C. Touchstone, Ed.) Wiley, New York (1982), pp. 261-277

Dévényi, T. and Pongor, S.: Hung. Sci. Instr. 50, 1-8 (1980)

Dévényi, T., Báti, J. and Fábián, F.: Acta Biochim. Biophys.
    Acad. Sci. Hung. 6, 133-138 (1971ə)

Dévényi, T., Hazai, I., Ferenczi, S. and Báti, J.: Acta Biochim.
    Biophys. Acad. Sci. Hung. 6, 385-388 (1971b)

Dévényi, T., Báti, J., Kovács, J. and Kiss, P.: Acta Biochim.
    Biophys. Acad. Sci. Hung. 7, 237-239 (1972)

Dévényi, T., Pongor, S., Pentzi, E. and Sirokmán, F.: Növény-
    termelés 25, 291-304 (1976)

Dévényi, T., Pongor, S., Váradi, A. and Axmann, H.: Int. Atomic.
    Energy Agency, Vienna, IAEA-RC-57/58 (1978) pp. 349-354

Dusha, I. and Dénes, G.: Anal. Biochem. 81, 247-250 (1977)

Elődi, P. and Karsai, T.: J. Liquid. Chromatog. 3, 809-831
    (1980)

Fábián, F.: Acta Biochim. Biophys. Acad. Sci. Hung. 12, 31-36
    (1977)

Ferenczi, S., Báti, J. and Dévényi, T.: Acta Biochim. Biophys.
    Acad. Sci. Hung., 6, 123-128 (1971)

Ferenczi, S. and Dévényi, T.: Acta Biochim. Biophys. Acad. Sci.
    Hung. 6, 389-391 (1971)

Frei, R.W.: in Progress in Thin-layer Chromatography and
    Related Methods (Niederwieser, A. and Pataki, Gy., Eds.),
    Vol. 2, Ann Arbor-Humprey Science Publ., Ann Arbor. (1971)
    pp. 1-61

Fuller, M.J.: in Chromatographic Review (M. Lederer, Ed.) Vol.
    14, Elsevier, Amsterdam (1971) pp. 45-76

Georgi, B., Niemann, E.G., Brock, R.D. and Axmann, H.: Int.
    Atomic Energy Agency, Vienna, IAEA-SM 230/81 1, 311-339
    (1979)

Gocan, S., Nascu, H., Marutoiu, C., Ursu, E. and Liteanu, C.:
    Chem. Abstr. 92, 149-197b (1980)

Grieg, C.G. and Leaback, D.H.: Nature 188, 310-311 (1960)

Hannig, K. and Wirth, H.: Z. Anal. Chem. 243, 522-530 (1968)

Hazai, I., Zoltán, S., Salát, J., Ferenczi, S. and Dévényi, T.:
    J. Chromatogr. 102, 245-249 (1974)

Heftmann, E., Ed.: Chromatography, Van Nostrand Reinhold Company,
    New York, 3rd Ed. (1975)

Higgens, P.E. and Kastner, R.E.: Antimicrob. Ag. Chemother.
    324-331 (1964)

Himoe, A. and Rinne, R.W.: Anal. Biochem. 88, 634-637 (1978)

Hrabák, A. and Ferenczi, S.: Acta Biochim. Biophys. Acad. Sci. Hung. 6, 383-384 (1971)

Hrabák, A.: J. Chromatogr. 84, 204-207 (1973)

Inonie, E. and Ogawa, E.: J. Chromatogr. 13, 536-541 (1964)

Issaq, H.I.: J. Liquid Chromatogr. 4, 955-975 (1981)

Issaq, H.I. and Dévényi, T.: J. Liquid. Chromatogr. 4, 2233-2241 (1981)

Izumiya, N. and Muraora, M.: J. Am. Chem. Soc. 91, 2391-2392 (1969)

Jones, D.A., Munford, J.G. and Gabbott, P.A.: Nature 247, 233-235 (1974)

Juhász, O. and Polyák, D.: Acta Agron. Acad. Sci. Hung. 25, 299-308 (1976)

Juhász, E., Ember, I., Tasnády, Zs., Karsai, T., Hauck, M. and Papp, Z.: Medical Genetics (Szabó, G., Papp, Z., Eds.) Excerpta Medica, Amsterdam (1977) pp. 707-712

Kádár-Pauncz, J.: J. Antibiotics (Japan) 25, 677-680 (1972)

Kádár-Pauncz, J.: J. HRC and CC, 1, 313 (1978)

Kádár-Pauncz, J.: J. HRC and CC, 4 287-291 (1981)

Kalász, H. and Nagy, J.: J. Liquid Chromatogr. 4, 985-1005 (1980)

Kalász, H., Nagy, J., Mincsovics, E. and Tyihák, E.: J. Liquid Chromatogr., 3, 845-855 (1980)

Karsai, T. and Elődi, P.: Acta Biochim. Biophys. Acad. Sci. Hung. 14, 123-132 (1979)

Karsai, T. and Elődi, P.: Mol. Cell. Biol., 43, 105-110 (1982)

Karsai, T., Ménes, A., Molnár, J. and Elődi, P.: Acta Biochim. Biophys. Acad. Sci. Hung. 13, 181-184 (1979a)

Karsai, T., Ménes, A., Molnár, J. and Elődi, P.: Acta Biochim. Biophys. Acad. Sci. Hung. 14, 133-142 (1979b)

Kerényi, Gy., Pataki, T. and Dévényi, T.: Hung. Pat. No. 170-287 (1976) U.S. Pat. 4.197.012 (1980)

Kirchner, J.G., Miller, J.M. and Keller, G.J.: Anal. Chem. 23, 420-425 (1951)

Kirnos, M.D., Khudyakov, I.Y., Alexandrushkina, N.I. and Vanyushin, B.F.: Nature 270, 369-370 (1977)

Kisfaludy, L., Löw, M. and Dévényi, T.: Acta Biochim. Biophys.
    Acad. Sci. Hung. 6, 393-403 (1971)

Kiss, P. and Kovács, J.: J. Liquid Chromatogr., 5, 1531-1540
    (1982)

Kovács, J.: Acta Paediatr. Acad. Sci. Hung. 14, 165-169 (1973)

Kovács, J.: Wiss. Zschr. Ernst-Moritz-Arndt Univ. 24 41-53
    (1975)

Kovács, J.: Acta Biochim. Biophys. Acad. Sci. Hung. 14,
    119-121 (1979)

Kovács, J.: Magyar Pediater 15, 225-228 (1981)

Kovács, J. and Kiss, P.: Acta Paediatr. Acad. Sci. Hung. 19,
    137-143 (1978)

Lederer, M.: J. Chromatogr. 144, 275-277 (1977)

Lepri, L., Desideri, P.G. and Coas, V.: J. Chromatogr. 79,
    129-131 (1973)

Lepri, L., Desideri, P.G. and Lepori, M.: J. Chromatogr. 116,
    131-140 (1976a)

Lepri, L., Desideri, P.G. and Lepori, M.: J. Chromatogr. 123,
    175-184 (1976b)

Lepri, L., Desideri, P.G., Landini, M. and Tanturli, G.:
    J. Chromatogr. 129, 239-248 (1976c)

Lepri, L., Desideri, P.G. and Heimler, D.: J. Chromatogr. 153,
    77-82 (1978)

Lepri, L., Desideri, P.G. and Helmer, D.: J. Chromatogr. 209,
    312-318

Lin, M.C.: in Methods in Enzymology (Collowice, S.P. and Kaplan,
    N.O., Eds.), Vol. 38, Academic Press, New York (1974), pp.
    125-135

Mangold, H.K., Schmid, H.H.O. and Stahl, E.: in Methods in Bio-
    chemical Analysis (D. Glick, Ed.), John Wiley and Sons
    Inc., New York (1964) pp. 394-451

Markovic, O. and Sajgó, M.: Acta Biochim. Biophys. Acad. Sci.
    Hung. 12, 45-48 (1977)

Moore, S. and Slein, W.H.: J. Biol. Chem. 192, 663-681 (1951)

Pataki, Gy.: in Advances in Chromatography (J.C. Giddings and
    R.A. Keller, Eds.), Vol. 7, Marcel Dekker Inc., New York
    (1968) pp. 47-86.

Pietrzyk, D.J., Rotsch, T.D. and Leuthauser, S.W.C.:
    J. Chromatogr. Science, 17, 555-561 (1979)

Pongor, S.: J. Liquid. Chromatogr. 5, 1583-1595 (1982)

Pongor, S. and Baitner, K.: Acta Biochim. Biophys. Acad. Sci.
    Hung. 15, 1-4 (1980)

Pongor, S., Kramer, J. and Ungár, E.: J. HRC and CC 3, 93-94
    (1980)

Pongor, S., Penczi, E., Sirokmán, F. and Dévényi, T.: Acta
    Biochim. Biophys. Acad. Sci. Hung. 11, 75-77 (1976)

Pongor, S., Váradi, A., Hevesi, G. and Glasschröder, Hung.
    Sci. Instr. 42., 1-6 (1978a)

Pongor, S., Kovács, J., Kiss, P. and Dévényi, T.: Acta Biochim.
    Biophys. Acad. Sci. Hung. 13, 117-121 (1978b)

Pongor, S., Kovács, J., Kiss, P. and Dévényi, T. : Acta
    Biochim. Biophys. Acad. Sci., 13, 123-126 (1978c)

Rabson, R., Burton, G.W., Hanna, W.W., Axmann, H. and Cross, B.:
    International Atomic Energy Agency, Vienna. IAEA-RC-57/32,
    279-291 (1978)

Randerath, K., Thin Layer Chromatography, Academic Press, New
    York, 2nd English ed. (1966)

Randerath, K. and Randerath, E.: in Methods in Enzymology (S.P.
    Colowien and N.O. Kaplan, Eds.), Vol. 121, Academic Press,
    New York (1967) pp. 323-347

Rao, M.V.P., Bhatia, C.R., Perschke, H., Axmann, H. and
    Hermelin, T.: Cereal. Res. Commun. 7, 293-301 (1979)

Sajgó, M. and Dévényi, T.: Acta Biochim Biophys. Acad. Sci.
    7, 233-236 (1972)

Sajgó, M. and Hajós, Gy.: FEBS Lett. 38, 341-344 (1974)

Senyavin, M.M., Shulga, V.A. and Rubinstein: Zhurn. Anal.
    Chim. 35, 2389-2393 (1980)

Sherma, J. and Fried, B.: Anal. Chem. 54, 45R-57R (1982)

Sigurbjörnsson, B., Brock, R.D. and Hermelin, T.: Int. Atomic
    Energy Agency, Vienna, IAEA-SM 230/86 1, 387-423 (1979)

Solti, M. And Telegdi, M.: Acta Biochim. Biophys. Acad. Sci.
    Hung. 7, 227-232 (1972)

Srivastava, S.P., Dua, V.K. and Chauhan, L.S.: Chromatographia
    12, 241-243 (1979)

Stahl, E.: Pharmacie 11, 633-637 (1956)

Stahl, E.: Thin-layer Chromatography. A Laboratory Handbook.
    Springer, New York, 2nd English ed. (1965)

Stahl, E. and Mongred, H.R.: in Chromatography (E. Heftmann,
    Ed.) Van Nostrand Reinhold Company, New York, 3rd edition
    (1975) pp. 164-188

Szabó, B., Beregszászi, Gy. and Schlammadinger, J.: Orv. Heti-
    lap 116, 3061-3063 (1975)

Szirtes, V., Pongor, S. and Penczi, E.: Növénytermelés 26,
    49-59 (1977)

Tasnády, Zs., Ember, I., Juhász, E., Karsai, T. and Elődi, P.:
    in Medical Genetics (G. Szabó and Z. Papp, Eds.) Excerpta
    Medica, Amsterdam (1977) pp. 517-521

Telegdi, M., Fábián, F., El-Sewedy, S.M. and Straub F.B.:
    Biochim. Biophys. Acta 429 860-869 (1976)

Tomasz, J.: J. Chromatogr. 84, 208-213 (1973)

Tomasz, J.: Acta Biochim. Biophys. Acad. Sci. Hung. 9, 87-88
    (1974a)

Tomasz, J.: J. Chromatogr. 101, 198-201 (1974b)

Tomasz, J.: Anal. Biochem. 68, 226-229 (1975)

Tomasz, J.: J. Chromatogr. 128, 304-308 (1976)

Tomasz, J.: J. Chromatogr. 169, 466-468 (1978a)

Tomasz, J.: Chromatogr. 12, 36-37 (1978b)

Tomasz, J.: Chromatogr. 13, 345-349 (1980a)

Tomasz, J.: Chromatogr. 13, 469-471 (1980b)

Tomasz, J. and Farkas, T.: J. Chromatogr. 107, 396-401 (1975)

Tomasz, J. and Simoncsits, A.: Chromatographia, 8, 348-349
    (1975)

Talmi, Y.: Anal. Chem. 47/7, 658A-670A (1975)

Tyihák, E., Mincsovics, E. and Kalász, H.: J. Chromatogr. 174,
    75-81 (1979)

Uskert, A., Néder, Á. and Kasztreimer, E.: Magy. Kém. Folyói.
    29, 333-334 (1973)

Váradi, A.: J. Chromatogr. 110, 166-170 (1975)

Váradi, A. and Pongor, S.: J. Chromatogr. 173, 419-424 (1979)

Varga Szabó, E.: Orv. Hetilap 117, 663-664 (1976)

Welling, G.W., Croen, G. and Beintema, J.J.: Biochem. J. 147,
    505-511 (1975)

Winkler, U. and Schön, W.J.: Int. Atomic Energy Agency, Vienna,
Austria, IAEA-SM-260/47 1, 343-350 (1979)

Vondrácek, M.: in Chromatograph (E. Heftmann, Ed.), Van Nostrand
Reinhold Company, New York (1975) pp. 815-840
(1975)

Walton, H.F.: in Chromatograph (Heftmann, E.,Ed.) Van Nostrand
Reinhold Company, New York (1975) pp. 344-362

Yasuda, K.: J. Chromatogr. 87 565-569 (1973)

SYSTEMATIC AMINO ACID AND GLUTATHIONE ANALYSIS BY OVERPRESSURED THIN-LAYER CHROMATOGRAPHY

BÉLA POLYÁK, MAGDOLNA ÁBRAHÁM and LÁSZLÓ BOROSS

Department of Biochemistry, József Attila University, H-6726 Szeged, Közép fasor 52, Hungary

ABSTRACT

A new method has been developed for the systematic separation of amino acids: overpressured thin-layer chromatography. Combining the favourable properties of two well established solvent systems for amino acids, that of butanol - glacial acetic acid - water and of phenol - water, the desired amino acids forming proteins could be separated, with the change of the alcoholic or phenolic components. In this way besides the amino acids, different "rare" amino acids and glutathione could also be separated from biological samples. The method affords a possibility for rapid, routine analysis in laboratory practice.

INTRODUCTION

There are several well established chromatographic methods for amino acid analysis. Most of them are performed in normal chamber, developed in different solvent systems and on different plates /Pataki 1966/.

Determination of the "rare" amino acids, such as hydroxyproline and hydroxylysine was reported in our previous paper /Buzás et al. 1980/.

By the introduction of overpressured thin-layer chromatography /OPTLC/ by Tyihák et al. /1978, 1980/, wide application of high viscosity solvent systems became possible.

The oxidized and reduced forms of glutathione from an amino acid mixture were separated in phenol – water solvent system containing sodium dodecylsulfate /SDS/ /Ábrahám et al. 1983/. Good separation was obtained for the amino acids eluting after alanine, but apolar and aromatic amino acids remained together under the front in two groups.

Tyihák et al. /1981/ investigated the separation of amino acids in butanol – glacial acetic acid – water /4 : 1 : 1/ system performed in an overpressured ultra-micro chamber but the relative migration distances were not given. According to their chromatrographic behaviour the amino acids could be divided into four well-defined groups:

Group 1: basic amino acids　·
Group 2: sulfur-containing amino acids
Group 3: amino acids containing hydroxyl groups
Group 4: apolar amino acids.

Within one group the individual components could not be separated well from each other.

On the basis of the work mentioned above, we combined the two solvent systems for the purpose of the total analysis of amino acids and glutathione. This could be achieved in a single development, by changing the ratio of the different components in the solvent systems.

MATERIALS AND METHODS

Chemicals for general use were of analytical grade and purchased from Reanal Factory of Laboratory Chemicals /Budapest, Hungary/. Amino acid calibration mixture was purchased from BIO-RAD Laboratories /Richmond, California, USA/; phenol was obtained from Loba Chemie /Wien, Austria/ and it was vacuum distilled from magnesium before use. The Kieselgel 60 F_{245} and HPTLC chromatoplates were purchased from Merck /Darmstadt, FRG/.

The development of sheets was performed in Chrompres 10, pressurized ultramicro chamber purchased from Labor MIM /Budapest, Hungary/.

EXPERIMENTAL

Amino acids were applied on the plates in groups and their combinations are shown in Table I. They were dissolved in a /!/ mixture of methanol and 0.005 N HCl. The start strips contained 2 μg of amino acids for each component.

Table I. Composition of the amino acid groups investigated by OPTLC

Groups applied as a mixture						
1.	2.	3.	4.	5.	6.	7.
Phe	Trp	Pro	Leu	Cys	Gl.am[***]	Cys$_2$
Val	Ile	Tyr	Ala	Asn	Nor. Val.	Hy-Pro
Thr	Gln	Glu	GSH[*]	Met	Orn	Hy-Lys
Asp	His	Gly	GSSG[**]	Arg		
	Ser		Lys			

[*] GSH = reduced glutathione
[**] GSSG = oxidized glutathione
[***] Gl.am = glucoseamine

The following solvent systems were employed /all proportions are v/v/:
/1/: n-Butanol-glacial acetic acid-water /4:1:1/
/2/: n-Butanol-glacial acetic acid-1 % SDS /4:1:1/
/3/: Phenol-n-butanol-i-butanol-glacial acetic acid water /40:20:4:10/
/4/: Phenol-n-butanol-i-butanol-glacial acetic acid-water /40:20:6:12/
/5/: Phenol-i-butanol-glacial acetic acid-water /60:40:5:20/

/6/: Phenol-i-butanol-n-propanol-water-glacial acetic
    acid /70:8:2:30:1,5/
/7/: Phenol-1% SDS /in water/-i-butanol-n-butanol-n-
    -propanol-glacial acetic acid /70:28:2:2:2:1,5/
/8/: Phenol-water-i-butanol-glacial acetic acid
    /70:27:10:1,5/

The plates were over-developed in every case, placing
filter paper strips on their top. The run took place in the
overpressured ultra-micro chamber using 12 bar standard pressure at 32°C. The flow rate of the solvent was 180 - 240 µl/min
for Kieselgel 60 F_{245} and 100 - 150 µl/min for the HPTLC
plates.

In order to remove the phenol from the layers they were
dried at 140°C in 0.9 kPa/m^3 vacuum for 20 minutes, then
sprayed with the following staining reagent:

50 ml solution A + 50 ml solution B + 2 ml p-collidine.
Solution A contained 2 g of ninhydrin in 100 ml acetone. Solution B contained 1 g of copper acetate dissolved in 245 ml of
deionized water, to which 5 ml of glacial acetic acid and 250
ml of acetone were added.

In order to identify the different amino acids in groups,
the staining reagent contained γ-collidine. In this way, the
spots had a coloured appearance. To fix the colours the layers
were covered with 1:2 paraffin oil - petrol ether /70°/.

Migration distances of amino acids and glutathione were
measured relative to alanine.

The reproducibility of the development highly depended on
the pH values, ion concentration and the protein content of
the samples.

RESULTS AND DISCUSSION

Repeating the experiments in butanol - glacial acetic
acid - water solvent system /first solvent/, performed by
Tyihák et al. /1981/, we completed their results with the relative mobility values of amino acids and glutathione /Table

II A/1/. The previously mentioned amino acid groups were observed, however, the individual components could only be separated in very low concentrations. Glutathione runs close to arginine. The spots of glutamate, alanine, threonine, and glutamine are touching each other and they could be separated only in very low, concentrations /under 0.5 µg/; the same is true for serine, glycine and aspartate. Phenylalanine and tryptophane migrated together, while leucine and isoleucine migrated near to each other. However, cysteine was well separated from methionine and the other amino acids.

Adding SDS to this solvent system /second solvent/, the order of the components has not changed /Table II A/2/. The migration distance of certain amino acids increased, while that of cysteine decreased.

It can be stated that the butanol - acetic acid - water system was not suitable for the total analysis of amino acids. The amino acids migrated far from the α-front. Data in Table II A/1,2 show a very good separation for the amino acids running above alanine. The apolar and aromatic amino acids run in the more apolar region of the solvent systems, with high migration distance. The relative mobility of these components was higher than 1.50, but the relative mobility of others was smaller than 1.00.

In the two-component phenol - water solvent system proper separation could be obtained for amino acids and glutathione running under alanine /Ábrahám et al. 1983/. Amino acids near the front migrated with similar mobility forming two groups. Taking into account these results we tried to combine the appropriate properties of the two previous solvent systems.

Results of these experiments are summarized in Table II A and B/3-8. The relative mobility for the third and fourth solvent systems containing butanol and phenol in 1:1 ratio is shown. In both systems, total separation was obtained for amino acids above alanine especially for tryptophane. The spot of proline was localized under valine. The basic amino acids showed a low migration distance, but aspartate ran together with them. Arginine left the basic group and was localized between glutamate and glycine.

Table II A. Migration distances relative to alanine. Development was performed on HPTLC plate

	1.	2.	3.	4.
Asp	0.761	0.676	0.114	0.273
Cys	0.304	0.338	0.085	0.136
Glu	0.956	0.946	0.443	0.523
Ser	0.812	0.824	0.314	0.568
Lys	0.116	0.243	0.186	0.182
Gly	0.783	0.798	0.628	0.636
Thr	0.971	0.987	0.771	0.795
His	0.188	0.284	0.228	0.318
Ala	1.000	1.000	1.000	1.000
Gln	0.978	0.703	0.657	0.750
Arg	0.231	0.338	0.543	0.125
Val	0.536	1.540	2.028	1.932
Tyr	1.884	1.824	2.200	2.068
Met	0.464	0.487	-	0.352
Pro	0.696	1.325	1.771	1.750
Ile	2.006	1.906	2.770	2.591
Leu	2.058	1.959	2.914	2.727
Phe	2.072	2.028	3.257	2.954
Trp	2.072	2.028	3.114	2.841
GSSG	0.203	0.310	0.000	0.068
GSH	0.623	0.365	0.057	-
Orn	0.116	0.243	0.114	0.114
Hy-Lys	-	0.229	-	0.091
Asn	0.580	0.635	0.514	0.523
Gl.am.	0.855	0.906	0.557	0.534
Hy-Pro.	-	0.757	-	1.045
Nor.Val	1.710	1.689	2.343	2.159

Table II B. Migration distances relative to alanine. Development was performed on HPTLC plate

	5.	6.	7.	8.
Asp	0.170	0.300	0.243	0.243
Cys	0.212	0.360	0.333	0.314
Glu	0.425	0.460	0.410	0.457
Ser	0.489	0.540	0.551	0.571
Lys	0.298	0.580	0.615	0.629
Gly	0.651	0.630	0.679	0.686
Thr	0.808	0.870	0.858	0.857
His	0.532	0.800	0.935	0.757
Ala	1.000	1.000	1.000	1.000
Gln	0.851	1.070	1.090	1.071
Arg	0.617	0.885	1.141	1.114
Val	1.872	1.650	1.538	1.543
Tyr	1.979	1.680	1.615	1.614
Met	2.149	1.770	1.731	1.700
Pro	1.872	1.810	1.756	1.728
Ile	2.340	1.840	1.783	1.779
Leu	2.425	1.870	1.846	1.836
Phe	2.638	1.990	1.897	1.900
Trp	2.553	2.010	1.890	1.900
GSSG	0.085	0.210	0.203	0.114
GSH	-	0.700	0.623	0.700
Orn	0.234	-	0.436	0.500
Hy-Lys	0.170	-	0.346	-
Asn	0.596	0.760	0.756	0.828
Gl.am.	0.468	-	0.628	0.729
Hy-Pro	1.149	-	1.256	-
Nor.Val	2.064	-	1.590	1.650

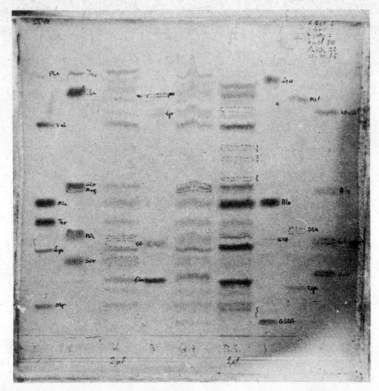

Fig. 1 Total amino acid and glutathione analysis from a
standard amino acid mixture /St +/ and a biological
sample /BS/ using the seventh solvent system. The
biological sample shows the free amino acid content
of intestine extracted from Cyprinus carpio. The
components marked with "U" represent unidentified
spots.

In the fifth system proline migrated between leucine and
tyrosine. The mobility of acidic amino acids decreased and the
basic amino acids were not grouped together.

In the sixth and seventh systems the ratio of alcohols was
decreased, while the relative amount of water was increased.
Table II B/6-7 and Fig. 1 show that these solvents are suitable
for total amino acid analysis. There was no marked limit between the distribution of polar and apolar amino acids, but it
was gradual. Tryptophane runs close to phenylalanine, but they
could be identified by polychromatic staining. /During the
acidic hydrolysis of proteins tryptophane degrades./

Reducing the relative amount of water in the solvent system, the separation of polar amino acids decreased /Table II B/8/.

Summarising our results, a new method has been developed for one-dimensional total analysis of amino acids with help of OPTLC. With the selection of suitable ratio of alcoholic and phenolic components in the solvent systems, any aminous acid can be separated. The introduced solvent systems can be employed not only for the rapid analysis of amino acids from protein hydrolysates but also for the determination of "rare" amino acids and glutathione from biological samples.

REFERENCES

1. G. Pataki : Dünnschichtchromatographie in der Aminosäure- und Peptid-Chemie. Berlin, Walter De Gruyter and Co. 1966.

2. Zs. Buzás, B. Polyák and L. Boross: Acta Biochim. Biophys. Acad. Sci. Hung. 15, 173. /1980/

3. E. Tyihák, H. Kalász, E. Mincsovics and J. Nagy: Proc. 17th Hung. Annu. Meet. Biochem., Kecskemét, 1977; C.A., 88, 15386. /1978/

4. E. Mincsovics, E. Tyihák and H. Kalász: J. Chromatogr. 191 /1980/ 293.

5. M. Ábrahám and B. Polyák: J. Liquid Chromatogr. /in press/

6. E. Tyihák, E. Mincsovics, H. Kalász and J. Nagy: J. Chromatogr. 211 45. /1981/

Chromatography, the State of the Art
H. Kalász and L.S. Ettre (Eds)

APPLICATION OF PRINCIPAL COMPONENT ANALYSIS TO COMPARE DIFFERENT THIN-LAYER CHROMATOGRAPHIC SYSTEMS

TIBOR CSERHÁTI, BARNA BORDÁS, MARGIT DÉVAI* and
ANNA KUSZMANN**

Plant Protection Institute of the Hungarian Academy of
Sciences, Budapest, Hungary
*Technical University of Budapest, Department of Applied
Chemistry, Budapest, Hungary
**Technical University of Budapest, Department of Inorganic
Chemistry, Budapest, Hungary

INTRODUCTION

The performance of a chromatographic system is generally characterized by the retention of solutes and/or by the ability to separate two hardly separable solutes. Therefore, these parameters have been also applied to compare different chromatographic systems. These parameters are undoubtedly very important from the practical point of view but they do not take into simultaneous consideration the behaviour of all solutes in the chromatographic systems to be compared. We think that the multivariate mathematical-statistical techniques such as principal component analysis, factor and cluster analysis are specially adequate to compare systems on the basis of the retention of all solutes (1, 2).

Up to now factor analysis was applied for deconvolution of two or more overlapping peaks in gas-liquid chromatography (3) or to determine the number of components under a single chromatographic peak (4).

Nearly all applications of factor analysis to chromatography have involved gas chromatographic data (5). We strongly advocate that the evaluation of new, reproducible and sensitive thin-layer chromatographic techniques such as high-performance (6) and overpressured thin-layer chromatography (7, 8, 9, 10) also require exact mathematical methods. To our knowledge the multivariate techniques have not been applied as yet for the evaluation of reversed-phase thin-layer chromatographic (RPTLC) systems.

The growing demand for new bioactive compounds accelerated the application of quantitative structure-activity relationship studies (QSAR) in drug design. Lipophilicity is one of the physicochemical parameters used most frequently in QSAR (11, 12). It can be determined by partition between water : n-octanol (log P value) or by chromatographic methods such as RPTLC and high-performance liquid chromatography. As pointed out by Tomlinson (13) chromatographic methods give values for the lipophilicity of bioactive compounds which correlate sometimes better with the biological data than the log P parameters. Furthermore only small amounts of substances are needed for the chromatographic methods and they need not   be of high purity.

For the determination of lipophilicity by RPTLC impregnated silica plates (14, 15, 16) or octadecylsilylated silica plates (17, 18) have been applied. Recent research indicates that the log P values correlate better with the $R_M^{\cdot}$ values determined on non-impregnated plates. The uncovered silanol groups exert some side effects even after impregnation, therefore this effect has also to be taken into consideration when determining the lipophilic properties (19, 20).

For QSAR studies (21) we determined the lipophilicity of some bioactive heterocyclic quaternary ammonium salts (22) under various conditions (23, 24). To compare the various RPTLC systems the data were evaluated by principal component analysis (PCA).

MATERIAL AND METHODS

The chemical structure of the heterocyclic quaternary ammonium   salts are shown in Fig. 1. Table I summarizes the TLC and RPTLC systems used while the R_f values obtained are listed in Table II. The modified silica was prepared at the Institute of Inorganic Chemistry of the Technical University Budapest (25). The detailed methodology is given in ref. 24.

In order to obtain maximal information from the data PCA was carried out in two different ways:

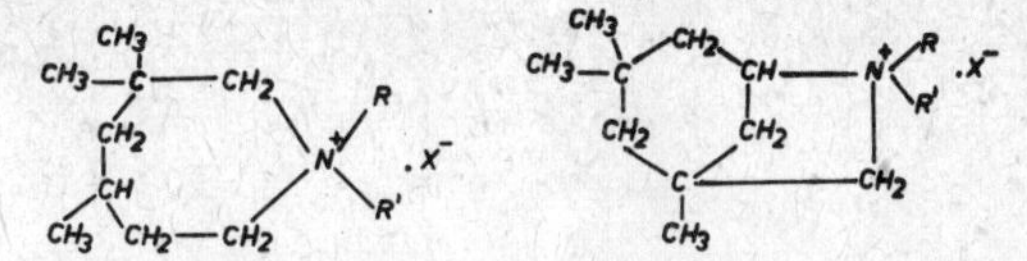

3,3,5-trimethyl-1-azo-cycloheptane derivatives

Compound number	R	R'	X^-
1	CH_3	CH_3	J
2	CH_3	H	J
3	CH_3CH_2	CH_3	J
4	$CH_2=CHCH_2$	CH_3	Br
5	$C_6H_5CH_2$	CH_3	Cl
6	$(CH_2)_4Br$	—	—

1,3,3-trimethyl-6-azo-bicyclo-(3.2.1.)-octane derivatives

Compound number		R'	X^-
7	CH_3	CH_3	J
8	CH_3	H	J
9	CH_3CH_3	H	J
10	$CH_2=CHCH_2$	H	Br

Fig. 1. Chemical structure of heterocyclic quaternary ammonium salts

PCA I.: R_f values as variables, compounds as observations;
PCA II.: Compounds as variables, R_f values as observations.
We prepared the linear mappings of the first two columns of PC loadings and variables. When the first two eigenvalues did not explain a sufficiently high percent of total variance, non-linear mappings of PC loadings and variables were carried out as well (26). We are well aware that the similar number of variables and observations could increase the probability of the occurrence of fortuitous correlations (27) but it is hardly probable that the same fortuitous correlations can occur simultaneously in both PCA. Therefore we assume that the conclusions drawn are sufficiently valid.

RESULTS AND DISCUSSION

The results of PCA I. are listed in Table III. The first four eigenvalues explain about 90% of the total variance, however a fairly high loading was also found in the fifth principal component. The systems containing ions have great loadings in the first component, that is, the ions modify strongly and uniformly the lipophilicity of heterocyclic quaternary ammonium salts, their effect overshadows the influence of support characteristics and that of the pH of the eluent: mainly

Table I. TLC and RPTLC systems to study the chromatographic behaviour of some heterocyclic quaternary ammonium salts (organic part of eluents always 50% methanol)

No.	Layer	Non-organic part of the eluent
1		water
2		0.1N NaOH
3		0.1N NaOH saturated with LiCl
4	silica gel impregnated	Britton-Robinson buffer of 6.00pH
5	with paraffin oil	Britton-Robinson buffer of 3.29pH
6		as No.4 + LiCl added to 0.61N
7		as No.4 + LiCl added to 1.22N
8		as No.4 + LiCl added to 2.43N
9		as No.4 + LiCl added to 4.86N
10		as No.4 + LiCl added to 9.73N
11		0.1 N HCl + LiCl added to 9.73N
12		as No.4 + LiCl added to 19.00N
13	silica gel	as No.1
14		as No.4
15		as No.12
16	silica gel with chloro-	as No.1
17	propyl groups covalently	as No.4
18	bound to the surface	as No.12

the ion environment determines the RPTLC behaviour of compounds.

This can be explained assuming that the ions can form ion-pair-like complexes with the quaternary ammonium groups modifying in this way their dissociation and lipophilicity. The ion-free systems show high and separate loadings in components 3, 4 and 5 underlining the special mode of retention of these systems.

From the graph of the nonlinear mapping of PC loadings (Fig. 2) the following conclusions can be drawn:

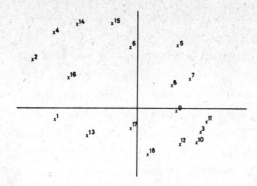

Fig. 2. Nonlinear mapping of PC loadings of PCA I.
        Number of iterations: 19. Error: 0.04.
        For symbols see Table I.

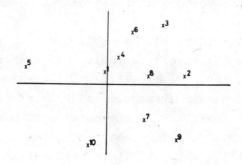

Fig. 3. Nonlinear mapping of PC variables of PCA I.
        Number of iterations: 33. Error: 0.05.
        For symbols see Fig. 1.

- The reversed-phase systems of high salt concentrations
  (points 3, 10, 11, 12, 18) form a group proving the
  similar effect of high salt concentrations independently
  of the fact that the lipophilic surface layer is un-
  bounded paraffin oil or covalently bounded chloropropyl
  group.
- Unmodified silica and reversed-phase stationay phases
  behave similarly at the neutral pH value (points 4 and
  14) supporting the theory that reversed-phase chromatography occurs in all cases when the eluent is more

hydrophilic than the support, i.e., it may occur also on "naked" silica (28),

- The ion-free systems, independently of the quality of the stationary phase form a loose cluster (points 1, 13, 16). In other words, in case of heterocyclic quaternary ammonium salts the influence of sorbent characteristics on their chromatographic behaviour are negligible.
- The non alkaline RPTLC systems containing lower salt con- centrations also form a loose group (points 5, 6, 7, 8, 9).
- No sorbent forms a cluster (points 1-12, 13-15 and 16-18 are far from each other). This means that the eluent has a higher influence on the chromatographic behaviour of heterocyclic quaternary ammonium salts than the sorbent.
- The chloropropyl modified silicas separate a little from all the other systems stressing their different reten- tion behaviour (points 16, 17, 18).

The two-dimensional nonlinear mapping of PC variables (Fig. 3) shows that the character of the heterocyclic ring does not determine the RPTLC and TLC behaviour of the com- pounds; compound 8 is below the 3,3,5-trimethyl-1-azo- cycloheptane derivatives (points 1-6).

Fig. 4. Linear mapping of PC loadings of PCA II.
        Error: 0.17. For symbols see Fig. 1.

Table II. $R_f \cdot 100$ values of some heterocyclic quaternary ammonium salts in different TLC and RPTLC systems (for symbols see Fig. 1 and Table I)

Compounds	Systems																	
	1	2	3	4	5	6	7	8	9	10	11	12	13	14	15	16	17	18
1	10	18	33	28	41	38	36	44	49	33	40	29	9	37	63	10	76	51
2	12	16	34	25	45	39	38	50	51	35	43	31	13	38	64	9	73	48
3	9	21	30	24	43	36	34	45	43	30	37	27	10	37	59	8	55	45
4	11	20	28	29	41	41	36	45	45	29	37	27	10	42	63	4	69	49
5	11	22	19	33	40	40	30	39	38	18	27	15	11	51	68	5	54	37
6	12	17	29	24	39	42	36	45	47	29	37	30	12	37	61	2	71	51
7	9	19	32	28	51	42	40	50	47	29	42	28	12	41	68	9	58	46
8	12	20	32	26	44	42	37	50	49	32	42	31	10	41	59	8	73	51
9	11	23	35	30	49	45	41	51	52	35	42	31	12	51	72	7	74	48
10	13	21	24	36	46	44	38	48	46	23	32	21	11	50	69	34	68	46

Table III. Results of PCA I.

	Eigenvalues	Sum of total variance explained %
1.	8.12	45.13
2.	4.67	71.05
3.	2.03	82.34
4.	1.26	89.33
5.	0.93	94.52

No. of system	Loadings				
	1	2	3	4	5
1	-0.02	0.39	-0.83	0.27	-0.02
2	-0.51	0.44	0.29	-0.48	-0.35
3	0.96	-0.02	0.22	-0.12	-0.05
4	-0.62	0.68	-0.18	-0.23	0.07
5	0.36	0.70	0.55	-0.04	0.24
6	0.09	0.85	-0.22	0.07	-0.28
7	0.76	0.61	0.11	-0.04	0.10
8	0.77	0.53	0.13	0.02	0.17
9	0.92	0.31	-0.14	-0.04	-0.03
10	0.95	-0.09	0.09	-0.12	-0.12
11	0.96	0.00	0.21	-0.04	-0.01
12	0.98	-0.13	0.02	-0.03	-0.11
13	0.21	0.44	0.17	0.85	-0.01
14	-0.54	0.77	0.00	-0.08	-0.29
15	-0.24	0.85	0.22	0.09	-0.14
16	-0.18	0.57	-0.29	-0.25	0.68
17	0.68	0.17	-0.62	-0.14	-0.20
18	0.85	-0.12	-0.40	-0.21	-0.01

Table IV. Results of PCA II.

	Eigenvalues	Sum of total variance explained %
1.	9.47	94.66
2.	0.32	97.82

No. of compounds	Loadings 1	2
1	0.98	0.12
2	0.98	0.16
3	0.99	0.06
4	0.99	0.03
5	0.92	-0.34
6	0.98	0.13
7	0.98	-0.01
8	0.99	0.13
9	0.99	0.01
10	0.91	-0.35

In the case of PCA II. the first eigenvalue explains about 95% of the total variance proving again the high similarity of the derivatives investigated (Table IV). All compounds have high loadings in the first component, while in the second component only the loadings of compounds 5, and 10 are of any importance. Because of the height of the first two eigenvalues the nonlinear mapping technique presumably would not have increased considerably the accuracy of the simple two-dimensional plotting of PC loadings and variables; therefore this calculation was not carried out. Due to their similar and high loadings in the first /PC component the compounds could be separated only according to the second PC component which explains only 3% of the total variance. Therefore the relative weight of separation is fairly low; however, Fig. 4 also supports our previous conclusions that the chemical structure of the heterocyclic ring does not determine the separation of these compounds.

Fig. 5. Linear mapping of PC variables of PCA II.
         Error: 0.30 For symbols see Table I.

From the two-dimensional linear mapping of the PC variables (Fig. 5) the same conclusions can be drawn as from
Fig. 3. The RPTLC systems with high LiCl concentrations form a
group as on the other map (points 3, 10, 11, 12). Silica and
impregnated silica behave very similarly in ion-free eluent
(points 1, 13). The non-alkaline RPTLC systems containing lower
LiCl concentrations also form a group (points 5, 6, 7, 8, 9).
       Summarizing our results we believe that the principal com-
ponent analysis proved to be a useful tool to compare different
thin-layer chromatographic systems taking simultaneously into
consideration the retention behaviour of all solutes.

SUMMARY

The thin-layer chromatographic behaviour of 10 heterocyclic quaternary ammonium salts was studied on silica gel, on
silica gel impregnated with paraffin oil and on silica gel with
surface-bonded chloropropyl groups using various water : methanol : LiCl mixtures buffered at different pH values. The data
set was evaluated by principal component analysis and visualized using non-linear mapping of the principal component
loadings and variables. It was established that the heterocyclic quaternary ammonium salts did not form clusters according to their chemical structure. The similarities between the

chromatographic systems are mainly governed by the eluent and not by the sorbent.

REFERENCES

1. K.V. Mardia, J.T. Kent and J.M. Bibby: Multivariate Analysis. Academic Press, New York, 1979.

2. E. Weber: Einführung in die Faktoranalyse. VEB Gustav Fischer Verlag, Jena, 1974.

3. D. MacNaughton, Jr., L.B. Rogers and G. Wernimont: Anal. Chem. 44 1421 (1977).

4. J.E. Davis, A. Shepard, N. Stanford and L.B. Rogers: Anal. Chem. 46 821 (1974).

5. E.R. Malinowsky, D.G. Howery: Factor Analysis in Chemistry. John Wiley and Sons, New York, 1980, p. 187.

6. R.E. Kaiser, ed.: Instrumental High Performance Thin-Layer Chromatography. Proceedings of the Second International Symposium. Interlaken, 1982. Institute of Chromatography, Bad Dürkheim.

7. E. Tyihák, E. Mincsovics and H. Kalász: J. Chromatogr. 174 75 (1979).

8. E. Mincsovics, E. Tyihák, H. Kalász and J. Nagy: J. Chromatogr. 191 293 (1980).

9. E. Tyihák, E. Mincsovics, H. Kalász and J. Nagy: J. Chromatogr. 211 45 (1981).

10. H. Kalász and J. Nagy: J. Liquid Chromatogr. 4 (6) 985 (1981).

11. A.J. Stuper, W.E. Brügger and P.C. Jurs: Computer Assisted Studies of Chemical Structure and Biological Function. John Wiley and Sons, New York, 1979.

12. R.F. Rekker: The Hydrophobic Fragmental Constant. Elsevier, Amsterdam, 1977.

13. E. Tomlinson: J. Chromatogr. 113 1 (1975).

14. C.B.C. Boyce and B.V. Milborrow: Nature (London) 208 537 (1965).

15. G.L. Biagi, A.M. Barbaro, M.F. Gamba and M.C. Guerra: J. Chromatogr. 41 371 (1969).

16. L. Ogierman and A. Silowiecki: J. High Resolut. Chromatogr./ Chromatogr. Commun. 4 357 (1981).

17. B. Rittich, M. Polster and O. Králik: J. Chromatogr. 197 43 (1980).

18. W. Butte, C. Fooken, R. Klussmann and D. Schuller: J. Chromatogr. 214 59 (1981).

19. M.C. Guerra, A.M. Barbaro, G. Cantelliforti, M.T. Foffani, G.L. Biagi, P.A. Borea and A. Fini: J. Chromatogr. 216 93 (1981).

20. W.F. van Giesen and L.H.M. Janssen: J. Chromatogr. 237 199 (1982).

21. Y.M. Darwish, T. Cserháti and Gy. Matolcsy: Acta Phytopath. Acad. Sci. Hung. 17 (1-2) 203 (1982).

22. Y.M. Darwish and Gy. Matolcsy: Z. f. angew. Entom. 91 252 (1981).

23. Y.M. Darwish, T. Cserháti and Gy. Matolcsy (1981): Proceedings of the 3rd Danube Symposium on Chromatography, Szeged, 1981, p. 239.

24. T. Cserháti, Y.M. Darwish and Gy. Matolcsy: J. Chromatogr. 241 223 (1982).

25. T. Cserháti, A. Kuszmann-Borbély and M. Dévai (1982): Magyar Kém. Folyóirat 88 451 (1982).

26. J.W. Sammon, Jr.: A Nonlinear Mapping for Data Structure Analysis. IEEE Transactions on Computers. C-18. (5) 401 (1969).

27. J.G. Topliss and R.P. Edwards: In J. Knoll: Chemical Structure-Biological Activity Relationships. Quantitative Approaches. Akadémiai Könyvkiadó, Budapest, 1980, p. 127.

28. E.B. Klaas, Cs. Horváth, W.R. Melander and A. Nahum: J. Chromatogr. <u>203</u> 65 (1981).

Chromatography, the State of the Art
H. Kalász and L.S. Ettre (Eds)

SOME PROBLEMS OF THIN-LAYER CHROMATOGRAPHY AS A PILOT TECHNIQUE FOR ADSORPTION LIQUID COLUMN CHROMATOGRAPHY

JAN K. ROZYLO, IRENA MALINOWSKA and MALGORZATA PONIEWAZ

Institute of Chemistry, M. Curie-Sklodowska University, Pl. M. Curie-Sklodowskiej 3, 20-031 Lublin, Poland

INTRODUCTION

Considering the wide application of chromatography it is necessary to find relations between the parameters obtained by different chromatographic techniques. The purpose of this study was to define some regularities between thin-layer and liquid column chromatography of some non-active organic substances in mixed binary solvent systems, i.e. between experimentally obtained log k values and theoretically calculated and in two different types of chambers (in the chambers of Stahl and Soczewinski) experimentally obtained R_M values. The advantage of thin-layer chromatography is that it can be used as a pilot technique for liquid column chromatography (1, 2). Theoretical calculations of the R_M values make it possible to use these data in HPLC. For this, it is necessary to define the correlations between R_M and the log k' values of chromatographed substances for a given mobile phase. These correlations change with changing mobile phase composition. From the study it appears that the simple theoretical dependence R_M = log k' does not exactly conform to experimental data. This discrepancy is due to the different conditions of the chromatographic process on a plate and in a column; i.e. the presence of a gas phase in Stahl's chamber, a different stationary phase profile on a plate and in a column, a different mobile phase flow-rate, etc.

The correlations between the log k' and R_M values obtained by using the chambers of Stahl and Soczewinski (3) for some substances such as naphthalene and its methyl derivatives and polycyclic aromatic hydrocarbons, in three different mobile phases have been shown.

In this paper the possibilities of using thin-layer chromatography as a pilot technique for liquid column chromatography are investigated.

MATERIALS AND METHODS

Experimental R_M values of the investigated substances were determined by ascending adsorption thin-layer chromatography. Silica gel 60 Merck was used as the adsorbent (4). The layer thickness was 0.25 mm. Plates were activated for 2 h at 135°C. The chromatograms were developed under thermostatic conditions at 21 ± 0.5 °C.

Pure isooctane, hexane, cyclohexane and chloroform, and their binary mixtures were used as the mobile phase. The chromatographed substances used were naphthalene and its methyl derivatives and polycyclic aromatic hydrocarbons. These substances were detected on the chromatograms in iodine vapour or UV light.

For liquid chromatographic experiments a Pye Unicam LC 20 Liquid Chromatograph was used. Merck LiChrosorb Si 60 of 5 μm particle diameter was used as an adsorbent The same substances and binary solvent mixtures were used as in TLC.

Experimental R_M values for the chromatographed substances were obtained using the chambers of Stahl and Soczewinski.

Oscik's thermodynamic theory of adsorption from multicomponent systems was used as the basis of our investigations. From the equation it is possible to calculate the theoretical R_M values for the studied substances (5-7):

$$R_{M1.2} = \varphi_1 \, \Delta R_{M1.2} + (\varphi_1^s - \varphi_1) \, (\Delta R_{M1.2} + A_z) + R_{M2} \qquad \text{eq. 1}$$

where $R_{M1.2}$, R_{M1}, R_{M2} are the R_M values of a given solute in a mixed mobile phase "1+2" and in the pure solvents "1" an "2" respectively; $\Delta R_{M1.2} = R_{M1} - R_{M2}$, is the difference of the R_M

616

values of a solute in pure solvents "1" and "2"; φ_1 is the
volume fraction of the stronger ("1") component of the mobile
phase; $\varphi_1^s - \varphi_1$ presents the excess adsorption of component "1"
of a mobile phase; the value A_z represents molecular interactions between the molecules of the chromatographed substances
and those of the mobile phase component. Experimental data for
equation 1 were obtained by using Stahl's chambers.

DISCUSSION AND CONCLUSIONS

Experimental and calculated data of the investigations are
given in Tables I-IV.

On the basis of the experiments and the calculated R_M
values and experimentally obtained log k' values of the studied
substances the correlations log k' = $f(R_M)$ have been plotted
for given mobile phase compositions (Figs. 1-9). From the
theoretical equation log k' = R_M it seems that these plots
ought to be straight lines with slopes equal to unity and that
they are independent of the mobile phase composition. From the
experimental data it follows that although the plots are
straight lines their slopes are not only different from unity
but are different for various mobile phase compositions as
well.

Figs. 1-6 show the graphical relationships between the experimentally measured log k' values and the experimentally obtained and theoretically calculated R_M values for hexane –
cyclohexane   and isooctane-hexane systems. These correlations
are straight lines having slopes of near unity. From the comparison of the slopes it seems that they change only insigni-
ficantly with the composition of the mobile phase. The variations in the slope values result rather from experimental
errors than from changes in the properties of the mobile phases
due to changes in their composition (Table V). From the plots
of Fig. 1a-6a obtained by using Stahl's chambers and Fig. 1c-6c
obtained by using Soczewinski's chambers it appears that the
relationships log k' = $f(R_M)$ are similar in the same mobile
phase systems (non-active type).

Table I. Capacity factor k′ values measured by liquid column chromatography on a column containing LiChrosorb Si 60, 5 μm

Mobile phase components		Volume fraction		Naphthalene	2-methyl-naphthalene	1,3-dimethyl-naphthalene	1,5-dimethyl-naphthalene	Anthracene	Pyrene	Chrysene	Fluoranthene	Phenanthrene
1	2	1	2									
Hexane	Isooctane	0.5	0.5	0.66	0.68	0.54	0.65	1.17	1.32	2.14	1.58	1.23
		0.7	0.3	0.58	0.60	0.66	0.65	1.10	1.20	2.14	1.55	1.17
		0.9	0.1	0.53	0.53	0.54	0.54	0.96	1.29	1.91	1.41	1.02
Hexane	Cyclohexane	0.5	0.5	0.37	0.31	0.31	0.31	0.49	0.56	0.81	0.63	0.53
		0.3	0.7	0.33	0.25	0.24	0.22	0.41	0.42	0.62	0.51	0.39
		0.1	0.9	0.17	0.18	0.15	0.15	0.37	0.47	0.60	0.49	0.37
Hexane	Chloroform	0.9	0.1	0.35	0.31	0.32	0.30	0.65	0.60	1.00	0.79	0.68
		0.7	0.3	0.25	0.18	0.18	0.19	0.33	0.37	0.46	0.41	0.32
		0.5	0.5	0.11	0.17	0.10	0.11	0.17	0.21	0.28	0.25	0.17

Table II. R_M values measured in Stahl's chamber on TLC plates prepared with silica gel 60.

Mobile phase components		Volume fraction		Naphthalene	2-methyl-naphthalene	1,3-dimethyl naphthalene	1,5-dimethyl-naphthalene	Anthracene	Pyrene	Chrysene	Fluoranthene	Phenanthrene
1	2	1	2									
Hexane	Isooctane	0.5	0.5	0.27	0.25	0.28	0.29	0.53	0.56	0.64	0.63	0.55
		0.7	0.3	0.28	0.23	0.29	0.28	0.50	0.55	0.66	0.60	0.53
		0.9	0.1	0.28	0.23	0.18	0.28	0.50	0.55	0.72	0.62	0.53
Hexane	Cyclohexane	0.5	0.5	0.36	0.39	0.43	0.43	0.57	0.63	0.75	0.70	0.60
		0.3	0.7	0.35	0.38	0.41	0.43	0.57	0.63	0.75	0.72	0.59
		0.1	0.9	0.35	0.38	0.41	0.43	0.60	0.63	0.75	0.74	0.60
Hexane	Chloroform	0.9	0.1	0.19	0.13	0.18	0.16	0.30	0.36	0.50	0.45	0.36
		0.7	0.3	0.00	0.00	-0.03	-0.02	0.09	0.10	0.19	0.17	0.12
		0.5	0.5	-0.08	-0.10	-0.06	-0.10	-0.06	-0.02	-0.04	-0.04	-0.04

Table III. R_M values calculated theoretically using equation 1.

Mobile phase components		Volume fraction		Naphthalene	2-methyl-naphthalene	1,3-dimethyl-naphthalene	1,5-dimethyl-naphthalene	Anthracene	Pyrene	Chrysene	Fluoranthene	Phenanthrene
1	2	1	2									
Hexane	Isooctane	0.5	0.5	0.27	0.25	0.28	0.28	0.52	0.55	0.64	0.63	0.54
		0.7	0.3	0.26	0.24	0.28	0.28	0.51	0.55	0.65	0.62	0.52
		0.9	0.1	0.18	0.23	0.29	0.28	0.52	0.55	0.72	0.62	0.51
Hexane	Cyclohexane	0.5	0.5	0.36	0.39	0.41	0.42	0.57	0.64	0.76	0.71	0.60
		0.3	0.7	0.35	0.38	0.41	0.42	0.57	0.63	0.75	0.71	0.60
		0.1	0.9	0.35	0.37	0.42	0.43	0.59	0.64	0.76	0.73	0.62
Hexane	Chloroform	0.9	0.1	0.14	0.10	0.16	0.13	0.28	0.34	0.44	0.37	0.32
		0.7	0.3	0.00	-0.03	-0.02	-0.02	0.05	0.11	0.13	0.12	0.11
		0.5	0.5	-0.08	-0.10	-0.06	-0.11	-0.06	-0.02	-0.03	-0.02	0.01

Table IV. R_M values measured in Soczewinski's chamber on TLC plates prepared with silica gel 60.

Mobile phase components 1	2	Volume fraction 1	2	Naphthalene	2-methyl-naphthalene	1,3-dimethyl-naphthalene	1,5-dimethyl-naphthalene	Anthracene	Pyrene	Chrysene	Fluoranthene	Phenanthrene
Hexane	Isooctane	0.5	0.5	-0.13	-0.08	-	-0.03	0.22	0.30	0.39	0.33	0.17
		0.7	0.3	0.00	-0.05	-0.01	-0.06	0.12	0.20	0.32	0.28	0.14
		0.9	0.1	-0.03	-0.03	-0.03	-0.05	0.20	0.19	0.31	0.32	0.14
Hexane	Cyclohexane	0.5	0.5	-0.13	-0.08	-	-0.04	0.11	0.15	0.22	0.23	0.09
		0.3	0.7	-0.10	-0.03	0.02	-0.05	0.16	0.28	0.34	0.16	0.15
		0.1	0.9	-0.13	-0.16	-0.12	-0.04	0.20	0.17	0.29	0.33	0.10
Hexane	Chloroform	0.9	0.1	-0.12	-0.17	-	-	-0.12	-0.05	-0.02	-0.04	-0.07
		0,7	0.3	-	-	-	-	-0.36	0.31	-0.22	-0.26	-0.29
		0.5	0.5	-	-	-	-	-0.64	-0.63	-0.51	-0.54	-

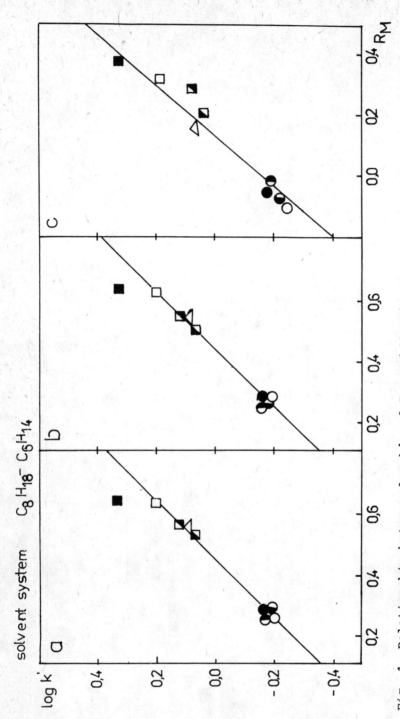

Fig. 1. Relationship between log k' and R_M. Mobile phase: isooctane-hexane, volume fraction of hexane, φ_1 = 0.5.

a = R_M values measured in Stahl's chamber; b = theoretically calculated R_M values; c = R_M values measured in Soczewinski's chamber.

Chromatographed substances: o naphthalene, ● β-methylnapthalene, ◐ 1,3-dimethyl-naphthalene, ● 1,5-dimethylnaphthalene, ◪ anthracene, ◨ pyrene, ■ chrysene, □ fluoranthene, △ phenanthrene.

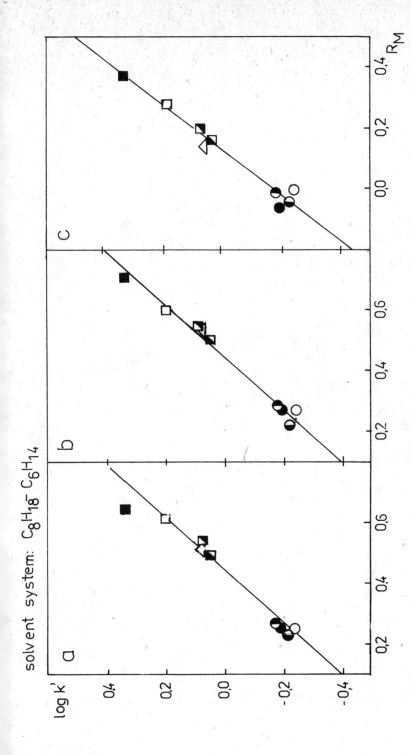

Fig. 2. Relationship between log k' and R_M. Mobile phase: isooctane-hexane; volume fraction of hexane, $\varphi_1 = 0.7$.

a = R_M values measured in Stahl's chamber; b theoretically calculated R_M values; c = R_M values measured in Soczewinski's chamber. For symbols see Fig. 1.

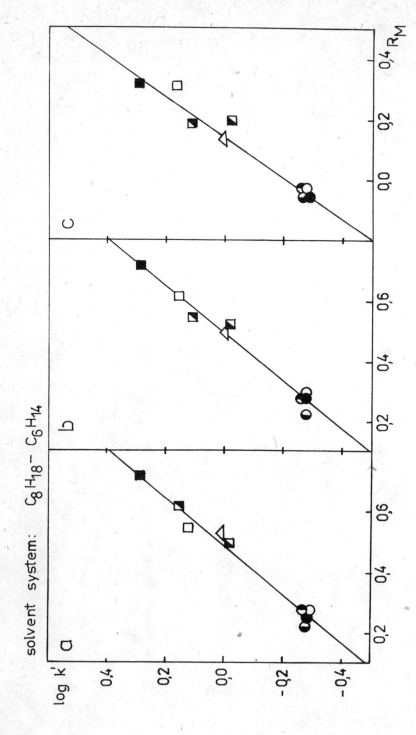

Fig. 3. Relationship between log k' and R_M. Mobile phase: isooctane-hexane; volume fraction of hexane, φ_1 = 0.9.

a = R_M values measured in Stahl's chamber; b = theoretically calculated R_M values; c = R_M values measured in Soczewinski's chamber. For symbols see Fig. 1.

solis system: $C_6H_{14} - C_6H_{12}$

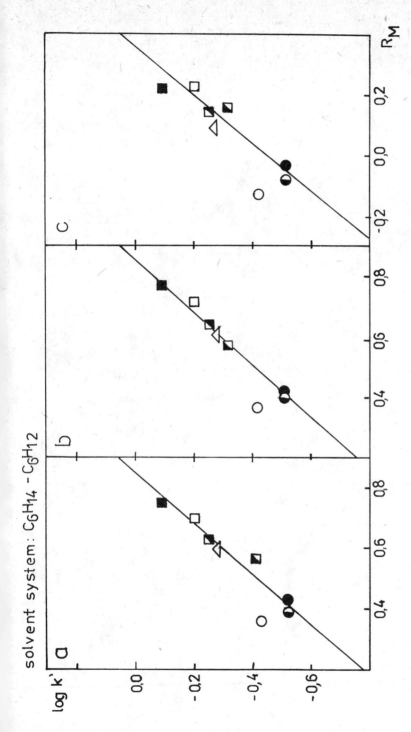

Fig. 4. Relationship between log k' and R_M. Mobile phase: cyclohexane-hexane; volume fraction of cyclohexane, φ_1 = 0.5.

a = R_M values measured in Stahl's chamber; b = theoretically calculated R_M values; c = R_M values measured in Soczewinski's chamber. For symbols see Fig. 1.

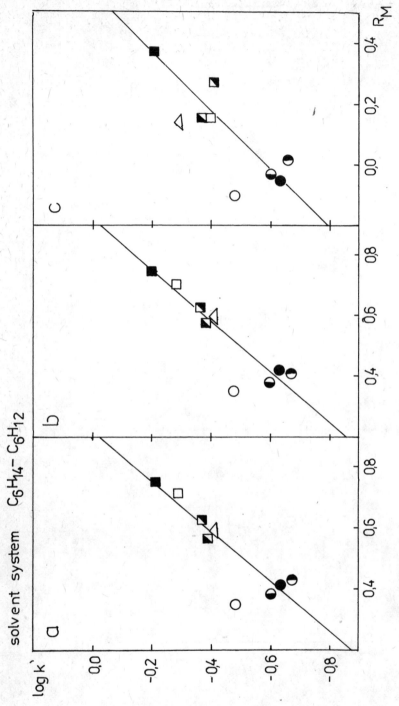

Fig. 5. Relationship between log k′ and R_M. Mobile phase: cyclohexane-hexane; volume fraction of cyclohexane, $\varphi_1 = 0.7$.

a = R_M values measured in Stahl's chamber; b = theoretically calculated R_M values; c = R_M values measured in Soczewinski's chamber. For symbols see Fig. 1.

solvent system: C_6H_{14} C_6H_{12}

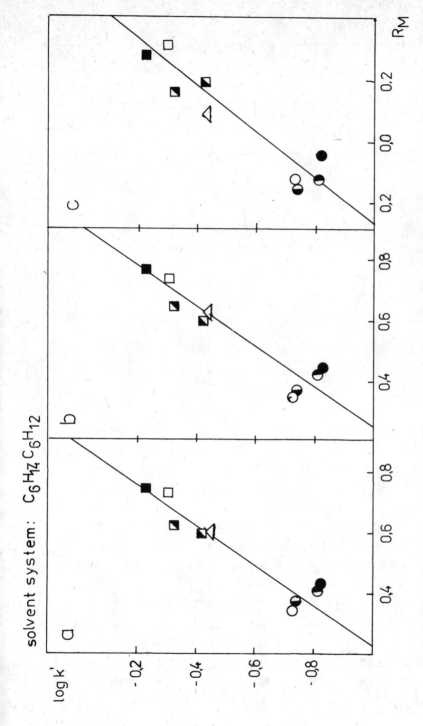

Fig. 6. Relationship between log k' and R_M. Mobile phase: cyclohexane–hexane; volume fraction of cyclohexane, $\varphi_1 = 0.9$.

a = R_M values measured in Stahl's chamber; b = theoretically calculated R_M values; c = R_M values measured in Soczewinski's chamber. For symbols see Fig. 1.

Table V. Slopes of the log k' = f(R_M) relationships

Mobile phase	1	Slopes*		
		a_1	a_2	a_3
isooctane-	0.5	1.05	1.10	1.02
hexane	0.7	1.10	1.10	1.04
hexane-	0.9	1.20	1.25	1.50
cyclohexane	0.5	1.20	1.15	1.25
	0.7	1.20	1.20	1.00
hexane-	0.1	1.45	1.50	2.70
chloroform	0.3	2.10	2.50	1.20
	0.5	6.10	6.50	1.27

* a_1 = slopes for the R_M values obtained in Stahl's chambers

a_2 = slopes for theoretically calculated R_M values

a_3 = slopes for the R_M values obtained in Soczewinski's chambers

The situation is different in the case of a hexane-chloroform system. Figs. 7-9 a, b show the correlations between log k' and the R_M values obtained by using Stahl's chambers and the theoretically calculated R_M values for selected mobile phase compositions for this system. Distinct changes in the slopes were observed with changing mobile phase concentration. The slopes for the straight lines log k' = f(R_M) obtained by the two TLC techniques and the R_M values calculated theoretically from equation 1 are presented in Table V. The straight lines become steep when the amount of chloroform ($_1$) in the mobile phase increases. In other words, the increase in the elution strength of the the mobile phase influences the shapes of straight lines describing the log k' = f(R_M) relationship.

The plots showing the log k' = f(R_M) relationship for the R_M values obtained experimentally by using Soczewinski's

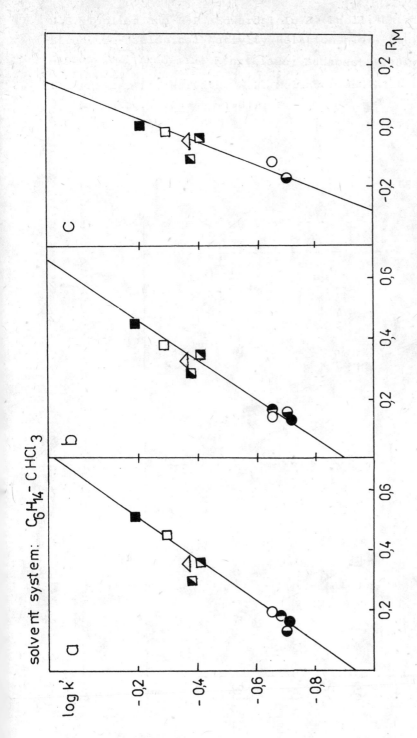

Fig. 7. Relationship between log k' and R_M. Mobile phase: hexane-chloroform; volume fraction of chloroform, $\varphi_1 = 0.1$.

a = R_M values measured in Stahl's chamber; b = theoretically calculated R_M values; c = R_M values measured in Soczewinski's chamber. For symbols see Fig. 1.

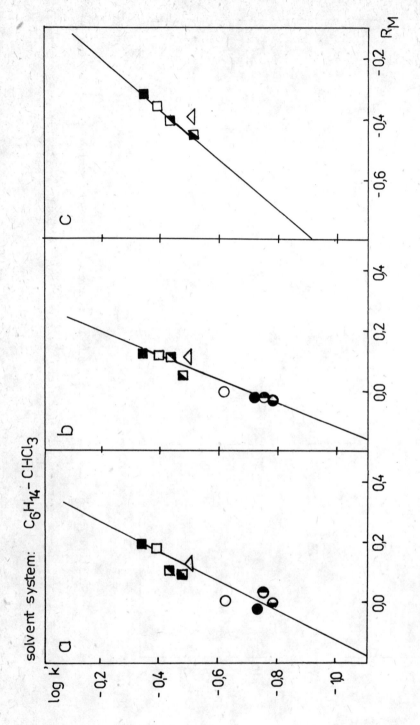

Fig. 8. Relationship between log k' and R_M. Mobile phase: hexane-chloroform; volume fraction of chloroform, $\varphi_1 = 0.3$.

a = R_M values measured in Stahl's chamber; b = theoretically calculated R_M values; c = R_M values measured in Soczewinski's chamber. For symbols see Fig. 1.

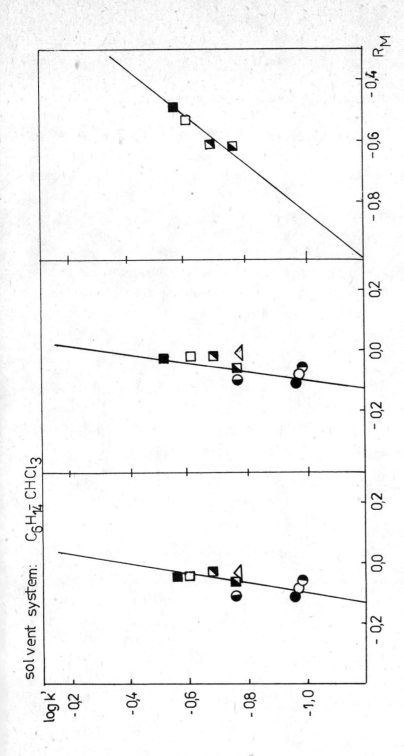

Fig. 9. Relationship between log k' and R_M. Mobile phase: hexane-chloroform; volume fraction of chloroform, $\varphi_1 = 0.5$.

a = R_M values measured in Stahl's chamber; b = theoretically calculated R_M values; c = R_M values measured in Soczewinski's chamber. For symbols see Fig. 1.

chambers (Fig. 7-9 c) have slopes near unity. It seems that the experimental data obtained by using Soczewinski's chambers are closer to the theoretical relationship $\log k' = R_M$. This means that the conditions of the chromatographic process in Soczewinski's chambers are close to the conditions in a chromatographic column.

From the data presented in this paper it appears that the correlation between $\log k'$ and the R_M values are straight lines and their shapes depend on the composition of the mobile phase and on the TLC technique used.

We can conclude that thin-layer chromatography can be used as a pilot technique for liquid column chromatography.

As it was noted above, it is better to use Soczewinski's chambers (especially for more active mobile phases) for this purpose, because in this case, the influence of the instrumental conditions on the slopes of the $\log k' = f(R_M)$ is negligible.

REFERENCES

1. W. Gołkiewicz, Chromatographia 14, 411 (1981)

2. W. Gołkiewicz, Chromatographia 14, 629 (1981)

3. E. Soczewiński, J. Chromatogr. 138, 443 (1977)

4. Catalog of Merck 1981

5. J. Ościk, Roczniki Chemii 31, 621 (1957), 34, 745 (1960)

6. J. Ościk, Przemysł Chemiczny 40, 279 (1961), 44, 129 (1965)

7. J. Ościk, in "Physical Adsorption from Multicomponent Phases" Nauka, Moscow, 1972 p. 132 (in Russian)

Chromatography, the State of the Art
H. Kalász and L.S. Ettre (Eds)

PECULIARITIES IN DETERMINING THE PARAMETERS OF ION-EXCHANGE THIN-LAYER-CHROMATOGRAPHIC SYSTEM IN THE PRESENCE OF COMPLEX FORMATION

M.P. VOLYNETS, R.N. RUBINSTEIN and L.P. KITAEVA

V.J. Vernaski Institute of Geochemistry and
Analytical Chemistry, USSR Academy of Sciences,
Moscow, USSR

1. THEORETICAL GROUNDS

The application of calculation methods used in ion-exchange column chromatography to ion-exchange thin-layer chromatography, in the case of linear isotherms and in the absence of additional chemical interactions has been discussed (1).

We have underlined the necessity to consider the influence of the isothermal form when calculating both the geometry of the spots and the relationship of the ion-exchange constants of the elements from the R_f value at sufficiently high ion concentration in solutions, the common situation in practice in both elution and frontal TLC (2). It has been shown that for elution chromatography, in the absence of complex formation (when $C_o/C_{01} << 1$) the R_f parameter can be expressed for the convex isotherms of the "tail" of the zone, or for the concave isotherms of the "head" by the following equation:

$$\frac{1}{R_f} - 1 = \frac{\chi}{\varepsilon}^z \left(\frac{a_o}{c_{01}}\right)^{z/z_1} \tag{1}$$

where a_o - full exchange capacity of the sorbent;

C_o and C_{01} - concentration of the exchanging and the exchanged ions respectively;

χ - constant describing the exchange of the two ions;

z, z_1 - charges of the respective ions;

ε - relative proportion of the void volume in the volume of the sorbent layer.

Let us discuss the case when monodentate ligand L is present in the mobile phase (MP), and forms a number of complexes $(ML)^{z-1}$, $(ML_2)^{z-2}$, $(ML_3)^{z-3}$, etc. with the M^{z+} ion. In the case of rapid establishment of ion equilibria in solution or between the phases, the mixture of these ions should behave in TLC as an individual substance with C_o concentration [where $C_o = C_M + C_{ML} + C_{ML_2} + C_{ML_3} + \ldots$ (2)] forming one zone in the chromatogram. Solution equilibria are described by the laws of the reacting masses:

$$\frac{C_{ML}}{C_M \cdot C_L} = k_1 \; ; \qquad \frac{C_{ML_2}}{C_M \cdot C_L^2} = k_2 \; ; \qquad \frac{C_{ML_3}}{C_M \cdot C_L^3} = k_3, \quad \text{etc.} \tag{3}$$

Let us introduce the function of complex formation:

$$F = \frac{C_o}{C_M} = 1 + k_1 C_L + k_2 C_L^2 + k_3 C_L^3 + \ldots \tag{4}$$

Let us mark the concentrations of the corresponding compounds in the solid phase as a_z, a_{z-1}, a_{z-2}, and their exchange constants for the exchanged ion as x_z, x_{z-1}, x_{z-2}. The concentration of the last complex, charged 1+ will be denoted as a_{01}. Using the Nikolsky equations, written in the form of concentration, we find the distribution coefficients (Γ_z) for every exchange ion:

$$\Gamma_z = \frac{a_z}{C_m} = (X_z \frac{a_{01}}{C_{01}})^z \; ; \quad \Gamma_{z-1} = \frac{a_{z-1}}{C_{M_L}} = (X_{z-1} \frac{a_{01}}{C_{01}})^{z-1}$$

$$\tag{5}$$

$$\Gamma_{z-m} = \frac{a_{z-m}}{C_{ML_{2-m}}} = (X_{z-m} \frac{a_{01}}{C_{01}})^{z-m}$$

where $m \leqslant z-1$.

The equivalent (summary) coefficient of Γ_3 distribution is

equal to
$$\Gamma_3 = \frac{a_z + a_{z-1} + a_{z-2} + \ldots a}{C_o} \tag{6}$$

or (in the case $C_o/C_{01} \ll 1$, when $a_{01} \to a_o$)

$$F \cdot \Gamma_3 = (X_z \frac{a_o}{C_{01}})^z + (X_{z-1} \frac{a_o}{C_{01}})^{z-1} \cdot k_1 C_L + (X_{z-2} \frac{a_o}{C_{01}})^{z-2} \cdot$$

$$\cdot k_2 \cdot C_L^2 + \ldots + X_1 \frac{a_o}{C_{01}} k_{z-1} C_L^{z-1} \tag{6a}$$

It is known that the R_f value can be expressed as

$$R_f = \frac{\varepsilon}{\Gamma_3 + \varepsilon} \tag{7}$$

As a conclusion we obtain:

$$(\frac{1}{R_f} - 1) F \cdot \varepsilon = \Gamma_3 \tag{7a}$$

The application of equations (1) and (7) for both finding the exchange constants of different complexes for the exchanging ion and the preliminary calculation of R_f will be illustrated below in more detail on concrete examples of the ion sorption of lanthanum and nickel from per chlorate and hydrochloric acid solutions

2. EXPERIMENTAL PART
Reagents and Equipment

Lanthanum solutions were prepared by dissolving certain amounts of (highly pure) lanthanum oxide and nickel chloride in 1 M $HClO_4$ or 1 M HCl. The concentrations of lanthanum and nickel varied within the ranges from $3.6 \cdot 10^{-3}$ to $9 \cdot 10^{-1}$ and from

$8 \cdot 10^{-2}$ to $1.7 \cdot 10^{-1}$ M respectively. Solutions of 1-4 M $HClO_4$ and of 1-10 M HCl served as the mobile phases. An 0.1% ($1.29 \cdot 10^{-3}$ M) aqueous arsenazo (III) solution was used to visualize the lanthanum zones, and a 0.1% ethanolic solution of rubeanic acid (dithiooxamide) to visualize the nickel zones.

Ready-made chromatographic plates (18x10 cm) with a layer of the cation exchange resin Dowex-50x8 ("Fixion", Hungary) in the Na-form were used. According to the manufacturer's information, the thin-layer contains commensurable amounts of silica gel together with the ion-exchange resin. According to specific measurements, the sorbent contains 53% of SiO_2, and the ignition lost in air amounts to about 42%.

Chromatography

Chromatographic experiments were carried out by ascending TLC, in chambers preliminarily saturated with the mobile phase vapours. 0.002 ml of the lanthanum or nickel solution were spotted on the plates and kept in the saturated chamber for a day before the actual chromatographic process. After the advancement of the mobile phase front to a height of 9 cm, the plate was taken out of the chamber, dried in air and sprinkled with the arsenazo (III) solution.

Determination of the layer thickness

The thickness of the sorbent layer (λ) was measured by a micrometer with ± 5 µm deviation on the difference of the thickness values obtained by measuring the layer thickness together with the bed and t .hickness of the bed after eliminating the sorption layer from it.

Determination of the full exchange capacity

The sorbent layer in the Na-form was taken from the 10x10 cm air-dried plates, weighed, and the sorbent was quantitatively transferred into a volumetric flask (its volume was 100 ml in the first case and 1000 ml in the second case). 0.5 M HCl solution was added up to the mark. The obtained solutions were kept for several days, while the sediment formed was shaken up periodically. It was found that in such conditions not more than 0.5% of Na remain in the sorbent. The Na content in the obtained solution was determined by flame photometry.

The exchange capacity for unit of sorbent volume (a_o) was calculated according to the following equation:

$$a_o = \frac{C_{Na^+} \cdot V}{M \cdot S \cdot \lambda} \qquad (\text{mg-equivalent/cm}^2);$$

where C_{Na^+} sodium concentration, mg/ml;

    M – g-equivalent weight of sodium;

    V – surface area of the sorbent layer, cm^2;

    λ – thickness of the layer, cm.

Determination of the value of ε

The value of ε was determined according to the volume of the liquid (aqueous NaCl solution), which served to fill the void volume between the particles of the air-dry sorbent. The sorbent layer in the Na-form was taken from two 10x10 cm plates (one of which was preliminarily saturated with a 0.5 M aqueous NaCl solution), transferred into a 200-ml volumetric flask which was then filled up to the mark by adding 0.5 M HCl. The Na content of these solutions was determined for a couple of days, and finally V_{NaCl} was calculated according to the difference of the Na contents. The value ε was calculated as $\varepsilon = \dfrac{V_{NaCl}}{V_{layer}}$, where $V_{layer} = S\lambda$.

Parallel to this determination, ε was also determined from the spot diameter (d), formed during spotting a drop of water of a certain volume (V_{drop}) on the sorbent layers (both air-dry and kept in the saturated chamber):

$$\varepsilon = \frac{4 \cdot V_{drop}}{\Pi \cdot d^2 \cdot \lambda}$$

The thickness of the layer was measured according to the methods described above.

3. DISCUSSION OF THE RESULTS
Determination of the layer thickness,
the full exchange capacity
and the value ε of the sorbent

According to the measurements the thickness of the sorbent layer varies from 110 to 190 μm from batch to batch. Sometimes fluctuations within one plate may reach 40 μm. This makes it necessary to control the layer thickness of each plate. A direct proportional dependence is established between the weight of the sorbent, taken from a certain area, and the thickness of the layer (Fig. 1) and the stability of the weight by volume of the sorbent, which was found to be equal to 0.5 g/cm^3. If the weight of the sorbent is known, the thickness of the layer can also be determined.

The full exchange capacity of the sorbent determined according to the methods given above, is equal to 0.69±0.02 mg equivalent/cm^3. For calculations we assume that the full exchange capacity of the sorbent, a_o = 0.69 mg-equivalent/cm^3. Due to the stability of the bulk weight this value does not depend on the thickness of the layer.

Results of determining the value of ε according to the drop volume and the diameter of the spot are given in Table I. Data for air-dry layers are confirmed by the ε measurement according to NaCl solution absorption. The value of ε measured from an air-dried layer was equal to 0.50±0.04, and the value of ε measured from a layer kept in the saturation chamber was equal to 0.34±0.04. Since all the chromatographic experiments are carried out on the plates kept in saturation chambers, the latter value was used in the subsequent calculations. It follows from the independency of the bulk weight from the layer thickness that the value of ε is also independent from the layer thickness.

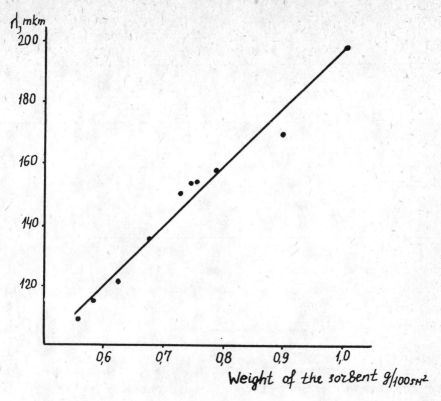

Fig. 1. Dependence of the thickness of the sorbent layer on its
        weight. Plate: "Fixion"

It follows from this discussion that the cited a_o and ε
values can be used for calculations according to eq. (1) and
similar values can be used for plates with layers of different
thickness.

Determination of the exchange constants

We assume that in the case of our experiment the lanthanum
- hydrogen exchange isotherm is convex. This is proven by the
data in Fig. 2 and Table II, which lead to the conclusion that
in elution TLC, in the area of the spot tail where $C_{La} \to 0$, $R_f x$
does not depend on the lanthanum concentration. Dependence of

$\dfrac{1}{R_f^x}$ on $\dfrac{1}{(C_{H^+})^3}$, as described by eq. (1), is given in Fig. 3.

Table I. Determination of the void volumes

Volume of the spotted solution, ml	H_2O						1 M HCl				
	I			I				II		
	d, mm	λ, μm	ε	d, mm	λ, μm	ε		d, mm	λ, μm	ε
0.002	5.55	160	0.48	5.10	190	0.53		6.55	165	0.36
0.003	6.49	190	0.49	5.92	190	0.54		-	-	-
0.004	7.18	190	0.51	6.85	190	0.57		9.31	170	0.34

I - plates with air-dried layer
II - plates kept in the vapours of the corresponding solutions
d - diameter of the spot (average value)
λ - thickness of the sorbent layer

For lanthanum solutions in $HClO_4$ eq. (1) is valid in the whole hydrogen concentration range investigated (plot 2). This fact proves the postulated model of ion change of La^{3+} for H^+ without complex formation. The value of $a_o^3 x_3^3/\varepsilon$, calculated according to Fig. 3 (plot 2), is equal to 1000 ± 200. Hence, assuming that $a_o = 0.69$ and $\varepsilon = 0.34$, we have $x_{La^{3+}/H^+} = 10 \pm 1$.

For the lanthanum - HCl system the points in Fig. 3 (curve 1) do not coincide with the plot. This can probably be explained by the presence of the chloride complexes $LaCl^{2+}$ and $LaCl_2^+$ in the system, together with La^{3+} ions; the existence of those complexes was not taken into consideration in the derivation of eq. (1).

Let us examine the effects of complex formation in more detail.

According to the literature (3, 4), the stability constants of the $LaCl^{2+}$ and $LaCl_2^+$ complexes are equal to:

$$k_1 = \frac{C_{LaCl^{2+}}}{C_{La^{3+}} \cdot C_{Cl^-}} = 2.89 \quad \text{and} \quad k_2 = \frac{C_{LaCl_2^+}}{C_{La^{3+}} \cdot C_{Cl^-}^2} = 0.38 \qquad (3a)$$

in the volume of the sorbent layer (ε)

	3 M HCl			Lanthanum solution in 1M-HCL (5 mgml)			
I		II		I		II	
d, mm	λ, ε μm	d, mm	λ, ε μm	d, mm	λ, ε μm	d, mm	λ, ε μm
5.12	190 0.51	6.93	168 0.32	5.00	190 0.54	5.94	190 0.38
5.88	190 0.58	-	-	5.79	190 0.6	-	
6.84	190 0.57	9.74	168 0.32	6.78	190 0.58	8.72	180 0.37

The total concentration of lanthanum ions in solution, usually determined analytically (e.g. by spectral methods), is equal to

$$C_{Lao} = C_{La^{3+}} + C_{LaCl^{2+}} + C_{LaCl_2^{\pm}} = C_{La^{3+}}(1 + K_1 C_{Cl^-} + K_2 C_{Cl^-}^2) \quad (2a)$$

The function of complex formation can be expressed as

$$F = 1 + K_1 C_{Cl^-} + K_2 C_{Cl^-}^2 \quad (4a)$$

Taking into consideration the fact that at $C_{Lao} \to 0$ $C_{H^+} = C_{Cl^-}$, eq. (4a) can be transformed in the following way:

$$F = 1 + K_1 C_{H^+} + K_2 C_{H^+}^2 \quad (4b)$$

As it is known, the part of each lanthanum ion (α) present in the solution is equal to:

$$\alpha_{La^{3+}} = \frac{C_{La^{3+}}}{C_{Lao}} = \frac{1}{F} \ ;$$

$$\alpha_{LaCl^{2+}} = \frac{C_{LaCl^{2+}}}{C_{Lao}} = \frac{K_1 C_{a-}}{F} = \frac{K_1 C_{H+}}{F} \ ;$$

$$\alpha_{LaCl_2^+} = \frac{C_{LaCl_2^+}}{C_{Lao}} = \frac{K_2 C_{a-}^2}{F} = \frac{K_2 C_{H+}^2}{F} \ .$$

| C_{La} mol/ℓ | HCl concentration in | | | | | | | | | |
| | 1.5 | | 1.8 | | 2.0 | | 2.24 | | 2.5 | |
	R_f^h	R_f^t	R_f^h	R_f^t	R_f^h	R_f^t	R_f^h	R_f^t	R_f^h	R_f^t
$3.6\cdot10^{-3}$	4.7	2.3	9.3	5	10.4	6.2	15	11	16	12
$3.6\cdot10^{-2}$	11	2.5	15.6	5.8	17	7.1	23	11	25	12
$7.2\cdot10^{-2}$	14	2.8	17.9	5.8	19	7.1	26	11	29	12
$3.6\cdot10^{-1}$	26	2	28.5	5.3	33	7.0	40	12	43	13
9.10^{-1}	35	1.7	38	5	41	6.7	49	12	54	13

*Each value represents the average value $R_f \cdot 100$, obtained from four parallel determinations.

The dependence of the various lanthanum forms on the HCl concentration is shown graphically in Fig. 4. As seen, even at $C_{HCl} < 1$ M all three ion-forms are present in the solution.

Let us mark the concentrations of each ion in the solid phase by: $a_1 = a_{LaCl_2^+}$; $a_2 = a_{LaCl^{2+}}$; $a_3 = a_{La^{3+}}$ (in M). Also let us denote the exchange constants of the corresponding ions as X_1, X_2, X_3. Then distribution coefficient of each lanthanum ion will be:

$$\Gamma_1 = \frac{a_1}{C_{LaCl_2^+}} = X_1 \left(\frac{a_{H^+}}{C_{H^+}}\right);$$

$$\Gamma_2 = \frac{a_2}{C_{LaCl^{2+}}} = \left(X_2 \frac{a_{H^+}}{C_{H^+}}\right)^2; \qquad (5a)$$

$$\Gamma_3 = \frac{a_3}{C_{La^{3+}}} = \left(X_3 \frac{a_{H^+}}{C_{H^+}}\right)^3$$

values on the lanthanum concentration

HCl in the mobile phase*

| the mobile phase M | | | | | | | | | |
| 2.7 | | 3.0 | | 3.2 | | 3.6 | | 4.0 | |
R_f^h	R_f^t	R_f^h	R_f^t	R_f^h	R_f^t	R_f^h	R_f^t	R_f^h	R_f^t
18	14	23	17	28	20	31	23	–	27
26	14	33	18	34	21	38	25	45	27
29	14	35	18.6	37	21	41	26	47	27
42	15	48	19	55	22	56	27	69	28
53	16	62	19	64	20	66	27	74	28

Table III Values of the exchange constants of different ion forms of lanthanum chlorides (a_o = 0.69 mg-eq./cm^3; ε = 0.34)

Constants and lanthanum ion form	Without considering the activity coefficient $(C_H)_{min}$=3.3 m Z_{min}=170	Considering the activity coefficient $(C_H^I)_{min}$=2.2 m Z_{min}=220
x_1 $(LaCl_2^+)$	7.6	20.2
x_2 $(LaCl^{2+})$	3.2	4.1
x_3 (La^{3+})	9.0	7.2

643

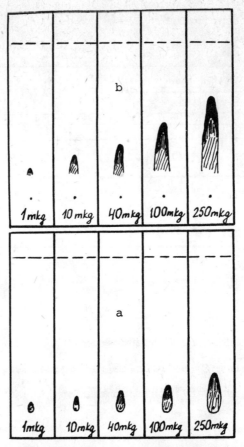

Fig. 2. Dependence of the form of lanthanum zone on its
content in the starting spot. Mobile phase:
a) 1M HCl; b) 3M HCl.

The equivalent (summary) distribution coefficient is equal to:

$$\Gamma_3 = \frac{\left(\dfrac{x_3 a_H^+}{C_H^+}\right)^3 + K_1 \left(\dfrac{x_2 a_H^+}{C_H^+}\right)^2 + K_2 x_1 a_H + C_H^+}{F} \qquad (6b)$$

where a_H^+ is the concentration of hydrogen ions in the ionite,
connected with a_o by the electroneutrality equation: $a_o = a_H +$
$+ 3a_3 + 2a_2 + a_1$.

When $a_1 + a_2 + a_3 = a_o$, which is true in our case for concentrations in the zone of the spot's tail, it is obvious that $a_H^+ \approx a_o$.
Therefore we shall always assume that $a_H^+ = a_o$.

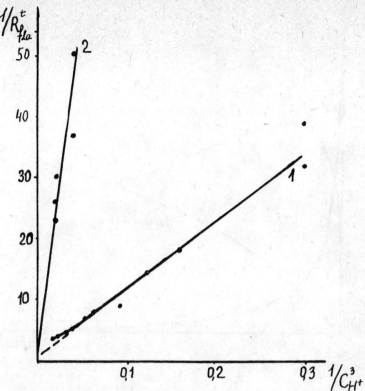

Fig. 3. Dependence of $\dfrac{1}{R^t_{fLa}}$ on $\dfrac{1}{c^3_{H^+}}$ for the chloride (1) and the perchlorate (2) systems

From equations (6) – (7) we have:

$$\left(\frac{1}{R^t_f} - 1\right) \cdot C_{H^+} = \frac{(x_3 a_o)^3}{\varepsilon C_{H^+}{}^2} + \frac{K_1 (x_2 a_o)^2}{\varepsilon} + K_2 \frac{(x_1 a_o C_H{}^2)}{\varepsilon} \qquad (8)$$

If we agree that

$$\left(\frac{x_3 a_o}{\varepsilon}\right)^3 = A \qquad (9); \qquad \frac{K_1 (x_2 a_o)^2}{\varepsilon} = D \qquad (10)$$

$$\frac{K_2 x_1 a_o}{\varepsilon} = B \qquad (11); \qquad \left(\frac{1}{R_f} - 1\right) F \cdot C_{H^+} = Z \qquad (12$$

645

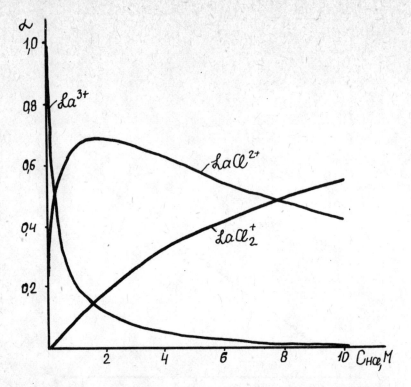

Fig. 4. Areas of existence of various ion forms of lanthanum
according to the HCl concentration in solution

then equation (8) transforms into

$$Z = \frac{A}{C_H^2{}^+} + B\ C_H^2{}^+ + D \tag{13}$$

Values of Z [see equations (4b) and (12)] are calculated
from experimental values R_f^t, obtained in elution TLC. The
corresponding plot describing the dependence of Z on C_H^+ is
given in Fig. 5, curve 1. It is clear from equation (13)   that
in the coordinates lg Z and lg C_H^+ the plot should be symmetrical and the slope of the asymptotes to it should be equal to
- 2 and +2, respectively, if $C_H^+ \to 0$ and $C_H^+ \to \infty$. The right part
of curve 1 does not correspond to these conditions and this

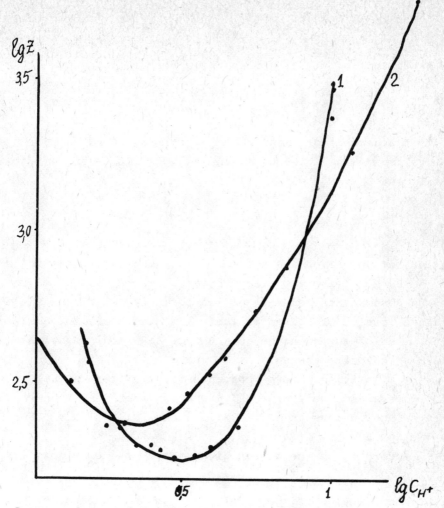

Fig. 5. Dependence of Z on C_H for the La–HCl system.
1 - without considering the activity coefficient;
2 - considering activity coefficients

is probably due to the the fact that the activity coefficients
were not taken into consideration. Fig. 5 also shows the
dependence of Z on $C_{H^+}^1 = C_{H^+} \gamma$ (curve 2), which meets the
demands listed above and proves this assumption. Here γ is the
value of the activity coefficient of the HCl solutions (5).

   Coefficients A, B, D in equation (13) can be connected to
the experimentally found two points on the curve in Fig. 5. We
find the first point Z_{min}; $(C_{H^+})_{min}$ by equalling $\frac{dZ}{dC_{H^+}}$ to zero;

hence $\frac{dz}{dC_{H^+}} = -\frac{2A}{(C_{H^+})^3_{min}} + 2B(C_{H^+}) \; min = 0 \quad (C_{H^+})min = (\frac{A}{B})^{\frac{1}{4}}.$

Substituting eq. (14) into equation (13) we get:

$$Z_{min} = 2\sqrt{AB} + D \tag{15}$$

The second point can be taken from anywhere on the experimental curve in Fig. 5 (11). For example, it is convenient to use point Z at $C_{H^+} = 1$

$$Z_{(1)} = A+B+D \tag{16}$$

After having solved the system of equations (14)-(16) in respect of A, B, D, it is quite easy to calculate the values of the exchange constants of various ion-forms of lanthanum. The obtained results are given in Table III. Value x_3, is equal to the value of the exchange constant of La^{3+} ion for hydrogen ion in a perchlorate system.

Similarly, exchange constants for different nickel ion-forms were also determined. It is known that in 3 M hydrochloric acid solution ions of Ni^{2+}, $NiCl^+$ and $NiCl_2^o$ ($k_1 = 0.56$; $k_2 = 0.90$ 6) are present.

According to the scheme given above [see equations (2) -
(7)] $C_{Ni^{2+}} + C_{NiCl^+} + C_{NiCl_2^o} = C_o$ (2b)

The function of the complex formation can be described as

$$f = 1 + k_1 \cdot C_{Cl^-} + k_2 \; C^2_{Cl^-} \tag{4}$$

Hence $C_{Ni^{2+}} = \frac{C_o}{F}$; $C_{NiCl^+} = \frac{C_o}{F} \cdot K_1 \; C_{Cl^-}$.

Ion distribution coefficients

$$\Gamma_1 = \frac{a_{NiCl^+}}{C_{NiCl^+}} = X_1 \frac{a_o}{C_{H^+}} ; \quad \Gamma_2 = \frac{a_{Ni^{2+}}}{C_{Ni^{2+}}} = (X_2 \frac{a_o}{C_{H^+}})^2 \tag{5}$$

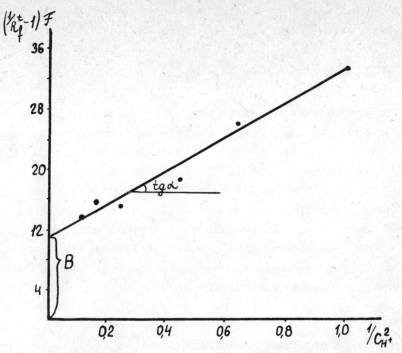

Fig. 6. Dependence of $(\frac{1}{R_f^t} - 1)$ F on $\frac{1}{C_{H^+}^2}$ for the chloride system Ni-HCl.

where χ_1 and χ_2 are constants of the exchange of ions of $[NiCl]^+$ and Ni^{2+} for hydrogen ion respectively

$$\Gamma_3 = \frac{a_{Ni^{2+}} + a_{NiCl^+}}{C_o} = \frac{(\chi_2 \frac{a_o}{C_{H^+}})^2 + \frac{\chi_1 a_o}{C_{H^+}} K_1 C_{Cl^-}}{F} \qquad (6b)$$

or (since $C_{Cl^-} = C_{H^+}$)

$$(\frac{1}{R_f} - 1) \; F = (\chi_2 \frac{a_o}{C_{H^+}})^2 \cdot \frac{1}{\epsilon} + \frac{\chi_1 a_o K_1}{\epsilon} \qquad (17)$$

Equation (17) is solved graphically for χ_1 and χ_2 (Fig. 6),

where $\quad tg\alpha = A = \frac{\chi_2^2 a_o^2}{\epsilon}$; $\quad B = \frac{\chi_1 a_o K_1}{\epsilon}$

Thus the found values of the exchange constants are equal to $X_1 = 9.24$; $X_2 = 4.08$.

Prediction of R_f

A reverse task was solved according to the suggested model: the value of R_f was calculated for the Ni^{2+} ions during their separation from lanthanum* on "Fixion" in perchlorate solutions, where complex formation of elements is absent.

The value of the Ni^2/H^+ exchange constant was found during the study of Ni sorption from hydrochloric acidic solutions as equal to $X_2 = 4.08$. This was used in the calculations. The R_f values of Ni^{2+} calculated and experimentally measured are given on Table IV. The results indicate the agreement of the calculation with the experimental data.

Table IV. Calculated and experimental values of $R_f^{Ni^{2+}}$ depending on the $HClO_4$ concentration

$HClO_4$ concentration in the mobile phase, M	R_f^t calculated	R_f^t determined
1	0.05	0.05
2	0.16	0.18
3	0.30	0.29
4	0.44	0.43

The calculated R_{fNi}^{2+} values in perchlorate media also agree with the experimentally measured R_{fNi}^{2+} values for nitric

---

* Such composition of the elements is observed during the analysis of several special filmic covers.

acidic solutions (where there is no complex formation either) obtained by Shulga (8), by using the same sorbent*.

Thus the precalculation of the R_f values permits to establish a priori that under the chosen conditions (in a 1-4 M - $HClO_4$ media), where the values of $R_{fNi}2+$ change in the range of 0.05÷0.43, separation of nickel from lanthanum, respecting the R_f range of 0÷0.03 (found experimentally), is possible. The calculation is experimentally proven by analyzing solutions obtained by dissolving lanthanum-nickel films.

CONCLUSIONS

The reliability of the suggested mathematical model for the determination of the ion exchange constants (in the absence of complex formation in solution) based on the R_f values of the "tail" of a zone obtained in elution TLC was demonstrated with the example of lanthanum and nickel sorption from perchlorate solutions on "Fixion 50x8" plates, by comparing calculated and experimentally measured results.

A new approach is suggested for the calculation of exchange constants of various ion forms of an element under sorption in the presence of complex formation in solution by the known dependencies of R_f on the ligand concentration and by the equilibrium thermodynamical characteristics of complex forms in solution.

ACKNOWLEDGEMENT

The authors express their sincere gratitude to Drs.N.S. Klassova, V.A. Pomytkina, and E.M. Sedykh for carrying out the analyses for the determination of the silicon and sodium content of the samples.

---

*The exchange constant value $x_{Ni}2+/H^+$ given by Shulga (9) does not coincide with any value determined in the present paper. The reason for this is that other values of the parameters of system-a_o and C were used by Shulga.

REFERENCES

1. Senyavin M.M., Shulga V.A and Rubinstein R.N. Zh. Analit. Khim. 35, (12) 2389-2393 (1980)

2. Rubinstein R.N., Volynets M.P. and Kitaeva L.P. Zh. Analit. Khim. 37, (8) 1370-1383 (1982)

3. Jatsimirsky K.B., Kostromina N.A., Sheka Z.A., Davydenko N.K., Kryss E.E. and Ermolenko V.I. Chemistry of the complexes of rare elements. Naukova dumka Kiev, 1966.

4. Kitaeva L.P. and Volynets M.P. Zh. Analit. Khim. 36, (8) 1490-1498 (1980)

5. Reference Book for Chemists, Vol. III.: Khimia Moscow, Leningrad 1964, p. 581.

6. Kivalo P. and Luoto R. Suomen Kemistileht, 30, (7) 163 (1957)

7. Shulga V.A. Ion-exchange thin-layer chromatography in the analysis of sewage. Thesis, USSR, 1980, p. 150.

8. ref. 7, p. 73.

9. ref. 7, pp. 76-78.

GAS CHROMATOGRAPHY

RAPID GAS-CHROMATOGRAPHIC DETERMINATION OF PHENYLACETIC ACID IN THE FERMENTATION BROTH OF BENZYL PENICILLIN PRODUCTION

DAN GH. MANCAS

Institute of Hygiene and Public Health, Str. V. Babes, 14, Jassy, 6600 - Romania

INTRODUCTION

During the fermentation process of the industrial production of benzyl penicillin the medium must contain a known quantity of phenylacetic acid, PAA. The monitoring of PAA levels requires rapid and specific analytical methods. Gas Chromatography, GC, was applied for the determination of free PAA during fermentation, using an acidic column to reduce peak tailing /1/. The problems associated with the GC analysis of free acids are well known /2, 3/. Their behaviour in the GC system is improved by derivatization, but the introduction of off-column derivatization is not satisfactory for the rapid monitoring of PAA in routine analysis.

Christophersen et al. obtained excellent results using the technique of on-column derivatization /4, 5/.

We studied this technique, and this paper presents a method for PAA, which is simpler and easier than the methods using direct analysis of free PAA.

MATERIALS AND METHOD

Reagents: Phenylacetic acid and benzoic acid were pure reagents /Carlo Erba, Italy/; benzene, puriss. /Fluka, Switzerland/, and N,O-bis/Trimethylsilyl/Trifluoroacetamide, BSTFA /Supelco, Bellefonte, U.S.A./.

655

Gas Chromatography: Fractovap Model 2350 /Carlo Erba, Italy/
gas chromatograph equipped with a flame ionization detector,
FID, was used. A Pyrex glass column /180 cm x 0.3 cm/ was packed with 4 % OV-101 on Chromosorb W-HP, 80-100 Mesh /Chrompack,
The Netherlands/ prepared by us, and the column ends were plugged
with quartz wool. Nitrogen was used as the carrier gas /1.5 at/
and hydrogen /0.6 at/ and air /1.1 at/ for the FID. Measurements were carried out at 124°C oven temperature, with the injection port and FID both heated to 200°C.

Procedure: After the centrifugation on the culture broth
for 2 min., 1 ml solution free of cells was placed into a glass-
stoppered tube, and acidified at pH=1 with sulphuric acid. The
solution was saturated with sodium chlcride and shaken for 1
min with 3 ml benzene containing 0.2 mg/ml benzoic acid /int.
std./. After a brief centrifugation, 1 µl benzene extract was
taken into a Hamilton syringe previously leaded with 2 µl BSTFA,
and injected into the gas chromatograph.

Quantitation: Calibration plot was obtained using standard
solutions of PAA with concentration in the range of 0.05 mg/ml
- 2 mg/ml benzene containing 0.5 mg/ml benzoic acid.

RESULTS AND DISCUSSION

A typical chromatogram obtained in the analysis of a culture broth sample contains the internal standard at 205 sec,
I, and PAA at 258 sec., II, and is shown in Fig. 1. No other
compounds were present in the sample at levels visible under
the analytical conditions used.
The calibration plot calculated with the peak heights was
linear in the experimental range. The equation calculated was:
$$Y = 0.72 \ X - 0.05$$
The coefficient of variation was 0.3 %.
BSTFA was used in this method for several reasons: quantitative silylation, volatility, and FID protection. However, Freon 113 was periodically injected. The volume of BSTFA required

for complete on-column silylation of PAA was 1 μl, but benzoic
acid reacted completely only with 2 μl reagents, in agreement
with the results of other authors for this acid /5/.

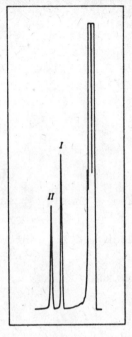

Figure 1.

The extraction of PAA followed. Niedermayer's procedure
/1/ but the volume of benzene was increased to 3 ml. This was
possible due to the better detection limit /to the nanogram
range/, which in fact was more than enough for the routine
analysis of the fermentation medium.

The described method has a number of advantages when compared with the methods which determine PAA in free form: 1/on-
ly 2 μl BSTFA is needed per sample; 2/ the separation column
has a long life, and is of general use, while in the analysis
of free PAA the column is specially treated, requiring sometimes conditioning with acids; also, the higher working temperatures reduce column life, and in addition, the packings are
affected by the humidity of the carrier gas; 3/ quantitation
is based on the internal standard technique; 4/ the GC system
is very stable under the working conditions; 5/ the method is
simple, and the maintenance of the GC system is not a problem;

6/ the analysis time is not longer even although the method included a derivatization step.

CONCLUSION

The proposed method is very simple in the industry. The absence of technical difficulties offers advantages even for untrained people.

REFERENCES

1. A.O. Niedermayer, Anal. Chem. <u>36</u>, 938 /1964/
2. D.M. Ottenstein, W.R. Szpina, J. Chromatogr., <u>91</u>, 119 /1974/.
3. D.V. McCalley, M. Cooke, C.A. Pennock, J. Chromatogr., <u>163</u>, 201 /1979/.
4. A.S. Christophersen, K.E. Rasmussen, J. Chromatogr., <u>168</u>, 216 /1979/.
5. A.S. Christophersen, K.E. Rasmussen, F. Tonnesen, J. Chromatogr,, <u>179</u>, 87 /1979/.

ANALYSIS OF THIODIGLYCOLIC ACID IN URINE BY CAPILLARY GAS CHROMATOGRAPHY

DAN GH. MANCAS

Institute of Hygiene and Public Health, 14. V. Babes Street
6600-Jassy, Romania

INTRODUCTION

Within the last years the identification of vinyl chloride /VC/ metabolites has been the subject of some important studies. Thiodiglycolic acid /TDGA/ was found /1, 2/ to be one of the main metabolites of VC monomer appearing in urine of humans and animals exposed to air containing different concentration of VC. Preliminary investigations made it clear that TDGA can also be a component of normal urine /3, 4/.

Gas chromatography /GC/ is a convenient technique for the analysis of TDGA, after its corresponding derivatization. GC with packed columns fails to provide an efficient method in situations when only small quantities of TDGA in urine have to be measured /4, 5/. This problem was overcome by using GC coupled with mass spectrometry /MS/, when less than 1 mg TDGA/l urine could be detected /3/. Watanabe et al.determined urinary TDGA by GC with packed column, but only after the urinary extract was fractionated by high-performance liquid chromatography /1/.

The more expensive GC-MS technique being only feasible in certain laboratories, the main purpose of this work was to develop a convenient method, in terms of simplicity and efficiency, for urinary TDGA, by using GC alone. Thanks to the high resolution of capillary columns the method is applicable for routine analysis of TDGA at urinary levels associated with occupational exposure to airborne VC concentrations near the allowable Threshold Limit Value /TLV/.

MATERIALS AND METHOD

Reagents: Ethyl acetate, ethanol, sodium chloride and anh. sodium sulphate were analytical reagents /"Reactivul", Bucharest, Romania/; ethyl acetate was distilled in glass before use. n--Hexane, analytical reagent /Merck, Darmstadt, G.F.R./ , boron trifluoride 47% in ether /Carlo Erba, Italy/, and thiodiglycolic acid, research grade, min. 99% /Serva, Heidelberg, G.F.R./.

Gas-Chromatography: The analyses were carried out on a Carlo Erba Model 2350 Instrument /Milan, Italy/, equipped with a flame ionization detector. The instrument was modified to accept capillary columns, being equipped with a home-made split-type injector, designed to assure linear splitting. It consisted of a large-bore silanized glass liner filled with silanized quartz wool, heated split-point, and a long, coiled tube in the went line. A glass wall-coated open-tubular /capillary/ column, 50m x 0.3 mm I.D., was used, coated with SE-52 gum phase /supplied by ICECHIM, Bucharest, Romania/.

The injector detector block temperature was $225^{\circ}C$, and the column oven was programmed as follows: initial hold 3 min at $108^{\circ}C$, then programmed to $155^{\circ}C$ at $1.1^{\circ}C/min.$; after the TDGA peak emerged, the temperature was quickly raised to $210^{\circ}C$ to purge the heavier compounds.

The inlet pressure of the carrier gas/nitrogen/ was 10 psig, controlled by a pressure regulator, and the gases at FID were hydrogen /5 psig/ and air /17 psig/.

Procedure: 10 ml urine adjusted at pH = 1 with HCl was saturated with NaCl and briefly centrifuged. A 5-ml aliquot was extracted three times with 15 ml ethyl acetate, by rotating the flask at 75 r.p.m., for 10 min. The combined extracts were treated with anh. Na_2SO_4, eliminating the water as a problem. Then, the extract was evaporated to dryness at $30-35^{\circ}C$ under vacuum.

To the dry residue 1.5 ml ethanol, and 0.4 ml BF_3-ether were added, and the mixture heated five minutes on a water bath /$90^{\circ}C$/. After cooling, it was extracted with 10 ml n-hexane, the extract washed with dist. water, and the hexane phase separated and concentrated to a volume suitable for GC analysis.

RESULTS AND DISCUSSION

Standard aliquots of TDGA corresponding to 2-30 mg/l urine were added to urine samples from the same batch. The samples were extracted and treated according to the procedure described above TDGA recovery from urine was 91%. The esterification yield was not considered.

For quantitative determinations the equation obtained was y = 0.347 x - 0.125. The calibration was based on the measuring of peak heights /absolute calibration/.

Identification of TDGA in urine extract was based on its retention time of 1104 sec., and co-chromatography on the sample with the standard added. Fig. 1 presents a chromatogram obtained for a urine sample containing 7 mg TDGA per liter. The detection limit of the described method is close to 1 mg TDGA per li-

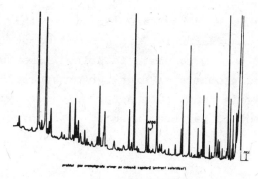

Figure 1

ter urine. Thus, the method is suitable for the monitoring of persons who are occupationally exposed to low levels of VC, even when urinary values of TDGA are close, but higher, than the normal values.

In spite of the complexity of chromatographic profile obtained for urine extracts, the resolution of the TDGA diester is good, pointing out the advantage of capillary columns for this analysis.

The derivatization of TDGA by esterification catalyzed by BF_3 gives stable derivatives. In addition, the extraction of the diethyl ester of TDGA with n-hexane acts as a purification step, at least partially. Other alternatives of derivatization, with diazomethane /3/, or silylation /4/, do not eliminate the compounds co-extracted from urine, and in the latter case the derivatives are unstable. The partial purification achieved in our case is not efficient enough to remove those compounds which interfere in TDGA analysis on packed columns /5/, but it gives a cleaner final extract.

Obviously, our results for the analysis of TDGA by capillary GC are better than those obtained using packed columns, and we used the proposed method for the routine analysis of TDGA in urine of persons who are occupationally exposed at low, medium, and high levels of VC in air, finding values in the range of 1 mg/l - 90 mg/l /7/.

CONCLUSION

Our work introduced high-resolution gas chromatography for the determination of urinary TDGA, pointing out that it is possible to detect this metabolite in urine without the utilization of mass spectrometry or high-performance liquid-chromatography, at levels found even after only low exposure to VC monomer. This method has been used by us in practices in the last years.

ACKNOWLEDGEMENTS

The author wants to thank Serva, Heidelberg, F.R.G. for a gift of thiodiglycolic acid, and Dr. W. Draminski from The Institute of Occupational Medicine Lódz, Poland, for valuable information and a sample of thiodiglycolic acid supplied for our first investigations.

REFERENCES

1. P.G. Watanabe, G.R. McGowan, P.J.Gehring, Toxicol. Appl. Pharmacol. 36, 339 /1976/.
2. S. Tarkowski, J.M. Wisniewska-Knypl, J. Klimczak, W. Dra-

minski, K. Wroblewska, J. Hyg. Epidemiol. Microbiol. Immunol. /Praha/, 24, 253, /1980/.

3. G. Müller, K. Norpoth, E. Kusters, K. Herweg, E. Versin, Int. Arch. Occup. Environ. Hlth., 41, 199 /1978/.

4. W. Draminsky, personal communication.

5. D. Gh. Mancas, Igiena /Bucharest/, 4, 337 /1981/.

6. G. Schomburg, H. Behlau, R. Dielmann, W. Weeke, H. Husmann, J. Chromatogr., 142, 87 /1977/.

7. D. Gh. Mancas, XVI Scientific Session of the Institute of Hygiene and Public Health Jassy, Romania, May 6-7, 1982.

GAS-LIQUID CHROMATOGRAPHIC METHOD FOR THE DETERMINATION OF SERUM CHOLESTEROL BY FLASH-HEATER SILYLATION

DAN GH. MANCAS

Institute of Hygiene and Public Health, 14. V. Babes Street, 6600-Jassy, Romania

INTRODUCTION

Clinical analytical methods must be rapid, specific, and simple. Although gas-liquid chromatography, GLC, possesses these characteristics, only a few GLC methods were reported for the analysis of serum cholesterol. In spite of the better gas chromatographic behaviour of derivatized cholesterol, the known methods determine it without derivatization /1, 2/. However, underivatized cholesterol sometimes gives tailing peaks, depending on the quality of the column packing /1, 3/. Also the GLC analysis of underivatized cholesterol requires high working temperatures which can affect the column packing and ordinary seals used in the instrument.

On the other hand, off-column derivatization is not advantageous in clinical analysis because the procedure becomes longer. Thanks to a new derivatization technique, "flash-heater derivatization" /4/, the GLC methods for cholesterol can be improved without prolonging the analysis time. Our studies pointed out that the slow elution of the derivatization agent leads to a tailing peak which reduces the sensitivity. Even at the GLC analysis of underivatized cholesterol the problem of interference due to the solvent peak imposed the choice of volatile solvent /petroleum ether 40°C/, although this is quite difficult to control at room temperature /2/.

The present paper describes an improved method for the GLC analysis of serum cholesterol based on flash-heater silylation, and a simple device which eliminates the reagent peak as a problem.

MATERIALS AND METHOD

Reagents: Methanol, toluene, and ethylene glycol were analytical reagents /"Reactivul", Bucharest, Romania/; isopropanol and n-hexane were of chromatographic grade /Loba Chemie, Wien, Austria/; Stigmasterol /Sigma, U.S.A./, and cholesterol /U.C.B., Belgium/ were used as received from the supplier. The silylation reagents were: Bis-/trimethylsilyl/trifluoroacetamide, BSTFA, and trimethylsilyldiethylamine, TMSDEA /Supelco, U.S.A./, hexamethyldisilazane, HMDS /Carlo Erba, Italy,/ and trimethylsilylimidazole, TMSI /Pierce, U.S.A./.

Gas-Liquid Chromatography: Carlo Erba Model 2350 instrument equipped with a flame ionization detector, FID, was used. The Pyrex glass column /180 cm x 0.3 cm/ was packed with 3% OV-17 on Gas-Chrom Q, 100-120 mesh /prepared by Supelco, U.S.A./. The column was part of an arrangement for venting the reagent, illustrated in Fig. 1. The vent line was localized at 10.5 cm

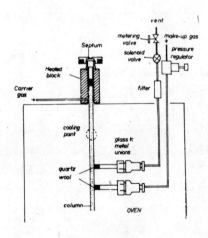

Figure 1

distance from the injector. The second gas line for the make-up gas was installed to maintain a constant flow through the column.

Argon was used as the carrier gas at 2.05 at; and the combustion gases were hydrogen /0.9 at/ and air /1.3 at/. Analyses were carried out at an oven temperature of 263°C, with the injector-detector block heated to 275°C. Periodically, Freon 113 was injected to reduce FID contamination, by combustion of the silyl derivatives.

The solenoid valve was activated at 80 sec before injection to open the vent line, and was closed 15 sec after injection.

Procedure: The extraction of free cholesterol from serum followed a method reported previously /2/. To a glass-stoppered tube containing 0.2 ml serum were added 1 ml methanol, 3 ml internal standard solution /0.314 mg stigmasterol/ml in a 1:1:1, v/v mixture of toluene, isopropanol, and methanol/, 3 ml 50% ethylene glycol, and 7 ml n-hexane. The tube was shaken by hand for 20 sec and the two layers separated. For the GLC analysis 2-4 μl from hexane layer were taken into a 10 μl Hamilton syringe previously loaded with 2 μl TMSI, and injected into the gas chromatograph.

RESULTS

The extraction procedure was identical to the method of Desager and Harvengt /90% recovery found with ^{3}H-cholesterol/ /2/. However, we used n-hexane instead of petroleum ether 40°C.

Figure 2 presents the analysis of free cholesterol in a serum sample which contained 34 mg/dl. The retention times for cholesterol /1/ and stigmasterol /2/ were 10 min, and 14 min, respectively.

Quantitative determinations were based on the internal standard technique, where stigmasterol is similar to cholesterol. The calibration plot was obtained with standard solutions at concentrations corresponding to the range of 700 - 30 mg cholesterol/dl serum. The plot given in Fig. 3 is linear in

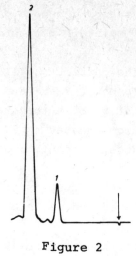

Figure 2

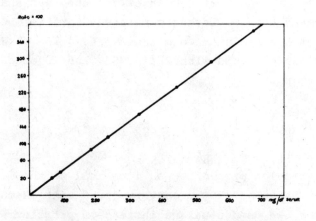

Figure 3

the experimental range. The same graph was obtained using two injection volumes, 2 and 4 μl. The volumes were varied due to several observations made in the literature concerning errors with the internal standard technique /5/.

The detection limit of the method is less than 8 mg/dl, and depends on the volume of the sample introduced into the GC.

DISCUSSION

The proposed method introduces two new aspects regarding the GLC determination of serum cholesterol: flash-heater /FH/ silylation, and the venting system. FH silylation has been used by other authors /4/ and different reagents were tried with various compounds. For the FH silylation of cholesterol and stigmasterol we investigated several reagents such as BSTFA, TMSDEA, HMDS and TMSI, but all of these failed to react satisfactorily, except TMSI which is specific for -OH groups /6/. Although TMSI is a high-boiling compound, the venting system solved this problem. The reaction with TMSI was rapid and complete, even with reagent volumes less than 1 μl. Increasing the volume of TMSI from 0.6 μl to several microliters results in no additional benefits. We used an excess of TMSI /2 μl/ to always assure complete silylation, and because this volume is easily vented.

The venting system was introduced with the purpose to eliminate the tailing of the solvent-reagent peak which obscures the peaks of interest at low attenuations. Such a system has also been used by other authors /7/. The difference in volatility between cholesterol and TMSI is large enough to enable the use of the venting system, even without the forming of a "cold spot" under our analytical conditions.

Our column and GLC parameters led to a relatively long retention of cholesterol. However, this can be easily improved to achieve retention times of 1-2 min. In this case, the importance of the venting system is obvious. This is illustrated in the chromatogram given in Fig. 4, where the column flow and temperature were slightly increased. If a shorter column containing less liquid phase would be used, the use of a cold spot to temporary trap the compounds of interest would be absolutely necessary.

Figure 4

The analysis of cholesterol by silylation had effect on the detection limit, as well as on the peak shape. A chromatogram obtained with a standard solution containing stigmasterol /0.773 mg/ml/ and cholesterol /0.747 mg/ml/ is shown in Fig. 5; the symmetrical shape of the peaks is obvious. Also, due to the derivatization, the analysis may be carried out using column packings which are not of top quality /1, 3/.

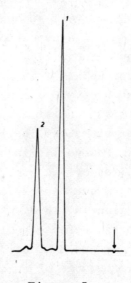

Figure 5

We applied the method for the determination of free serum cholesterol; however, taking into account the preparation procedure of serum reported by other authors /2, 8, 9/, this method is equally suitable for the analysis of total cholesterol, and cholesterol from low-density lipoproteins.

CONCLUSIONS

This work contributes to the improvement of the GLC analysis of serum cholesterol enabling to obtain a better detection limit, to use more convenient GLC parameters /e.g. relatively low temperatures/, and general-purpose column packings. We feel that finally, the presented results open the right way towards a very simple and rapid analysis of cholesterol in very small volumes of serum. Our future studies are directed to the optimization of this method, with the purpose of simultaneously increasing the speed of analysis and decreasing the sample volume.

ACKNOWLEDGEMENTS

I want to express our gratitude to Dr. C. Harvengt /Laboratoire de Pharmacothérapie de l'Université Catholique de Louvain, Brussels, Belgium/ for placing valuable information at our disposal and for a gift of stigmasterol.

Also, I thank Dr. Georgeta Mihail, head of our department, for the conditions created to achieve this work.

REFERENCES

1. J.L. Driscoll, D. Aubuchon, M. Descoteaux, H.F. Martin, Anal. Chem. _43_, 1196 /1971/
2. J.P. Desager, C. Harvengt, HRC and CC, _4_, 217 /1978/
3. K. Hammarstrand, E.J. Bonelli, "Derivative Formation in

Gas Chromatography", Varian Aerograph, Walnut Creek, California, 1968; p. 2.

4. A.S. Christophersen, K.E. Rasmussen, F. Tønnesen, J. Chromatogr. <u>179</u>, 87 /1979/

5. A. Shatkay, S. Flavian, Anal. Chem. <u>49</u>, 2222 /1977/

6. Handbook and General Catalog - 1983, p. 113, Pierce Eurochemie B.V. Beijerland, Holland.

7. F.F. Kaiser, C.W. Gehrke, R.W. Zumwalt and K.C. Kuo, J. Chromatogr., <u>94</u> 113 /1974/

8. M.F. Lopes-Virella, P. Stone, S. Ellis, J.A. Colwell, Clinical Chemistry, <u>23</u>, 882 /1977/

9. P. T. Kirch, "A routine method for the direct determination of low density lipoproteins-cholesterol", presented at The Second Joint Meeting of The Belgian, Dutch, German and British Societies for Clinical Chemistry, Newcastle upon Tyne, United Kingdom, April 19-22, 1983.

ISOLATION OF AND INVESTIGATION INTO THE ESTER FRACTION OF FRUIT FLAVOUR CONCENTRATES

F. BOROSS and M. TÓTH-MÁRKUS

Central Food Research Institute, Budapest, Hungary

Among the volatile constituents of fruit flavours representatives of nearly every class of organic compounds can be found. Their separation into components in one step is practically impossible even by the most modern separation methods. A given group of compounds previously isolated from the rest can be analyzed easier, by simpler means. Our paper reports on a method for separating esters which play a decisive role in fruit flavours.

For the isolation of esters, extraction and salting-out procedures are generally used /1, 2/. These methods are useful only for the concentration of esters and for their enrichment as related to the other classes of compounds but not for selective group separations. Palmer /3/ examined the column chromatographic isolation of esters on silica gel using Freon 11 as the eluent. The esters were not isolated from the aldehydes and ketones of lower polarity but separated well from other compound groups occurring in flavours. Murray and Stanley /4/ separated esters in micro amounts from a few oxo-compounds by ascending chromatography on a deactivated dry column packed with fine particle /HR/ silica gel.

In our work results similar to those of Palmer were obtained under preparative column chromatographic conditions, using n-pentane which is less polar and more easy to handle. Derivatization of aldehydes and ketones preceding column chromatography was not applicable because of the sensitivity of esters, however, these can be easily and quickly converted into alcohols by sodium borohydride reduction. Chromatography following the conversion of interfering compounds is useful for serial analysis but not recommended in the case of flavours of unknown composition. According to Hajós /5/ certain esters may partly decompose though this was not observed during our experiments.

A mild technique not affecting the ester composition was developed in our laboratory by coupling bisulphite adduct formation and column adsorption chromatography.

The equilibrium reaction of bisulphite adduct formation can be used for binding water-soluble lower carbonyls on ion-exchanger columns /6/. In the case of carbonyls with carbon numbers higher than four the reaction is not applicable under column partition chromatographic conditions, because with increasing chain length both the water solubility and the equilibrium constants abruptly decrease. Attempts were made to compensate the low reaction rate by decreasing the material flow. Partial coating of the silica gel having good adsorption properties with a saturated sodium bisulphite resulted in increasing the residence time of the carbonyl compounds to an extent permitting the reaction to take place.

By addition of 20-50 % saturated bisulphite solution to the silica gel, a new sorbent was obtained, active enough for the safe separation of esters from alcohols and containing a proper amount of bisulphite to bind the aldehydes and lower ketones under the usual column chromatographic conditions. Ketones with a carbon number higher than seven were not irreversibly bound but were retained and eluted together with the alcohols. 0.5 ml of the sample were applied to a 11 mm i.d. x 400 mm column, containing 20 g silica gel modified by bisulphite. At a 0.3 - 0.5 ml/min n-pentane eluent flow rate the esters were eluted within 35-200 ml, while the ketones and alcohols appeared only at 370 ml eluent volume.

After checking the new technique with test mixtures the investigation into the esters of a complex fruit flavour analyzed earlier by our research group was performed. The gas chromatograms of the original flavour concentrate and of its isolated esters are shown in Figure 1. Table 1 summarizes the quantitative changes in several characteristic compounds during the isolation of esters. These illustrate the applicability of the method. As a result of isolating the esters, the resolution of overlapping peaks was improved, thus the quantities of these could also be exactly determined. The isolation of esters simplifies separation to such an extent that for not too complex mixtures, the use of a packed chromatographic column is sufficient. As an illustration, Figure 2 shows the chromatograms of a simple fruit flavour concentrate and of its isolated esters.

Table I. Relative change in composition of a pear flavour concentrate rich in esters during the isolation /Peak No. 22 = 100 %/

Peak No.	Compound	Original flavour	Ester fraction
13	3-heptanone	0.1	0
14	amyl acetate	0.9	0.8
18	2-hexenal	1.1	0
20	ethyl hexanoate	2.7	2.7
21	1-pentanol	3.6	0
22	hexyl acetate	100	100
37	1-heptanol	2.3	0
38	octyl acetate	37	40
41	ethyl 4-oxy butenoate	19	0
43	1-octanol	9	0
48	ethyl decanoate	10.5	12
51	2-methylbutyric acid	12	0
59	ethyl decenoate	60	69

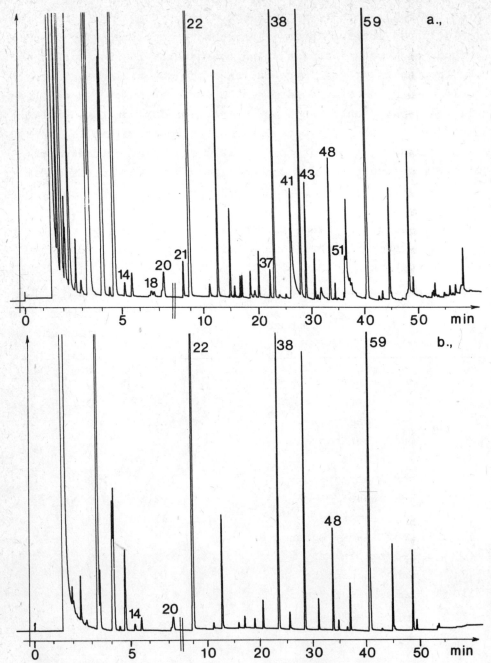

Figure 1. Chromatogram of a pear flavour concentrate
a, original flavour; b, ester fraction 0.3 mm
i.d. x 30 m glass WCOT Carbowax 20M column, initial temp. 60°C, program rate 2°C/min carrier
gas H_2, linear velocity 35 cm/sec.

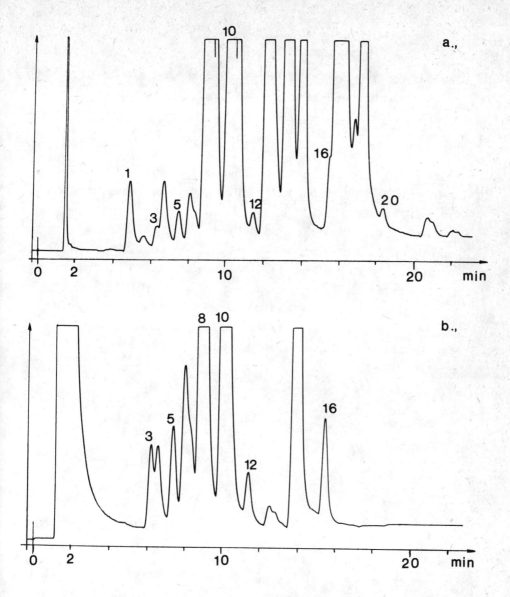

Figure 2. Chromatogram of an apple flavour concentrate
a/ original flavour; b, ester fraction 2.7 mm
i.d. x 4 m stainless steel column, 15 % FFAP
on Gas Chrom P 80-100 mesh. Initial temp. 60°C,
program rate 6°C/min carrier gas N_2, linear velocity 6 cm/sec.

ACKNOWLEDGEMENT

Thanks are due to dr. Tibor Tóth for his help in preparing the capillary chromatograms and in their evaluation.

REFERENCES

1. Hardy, P.J., J. Agric. Food Chem., 17, 656 /1969/
2. Woidich, H. and Pfannhauser, W., Deutsche Lebensm.-Rundschau, 74, 397 /1978/
3. Palmer, J.K., J. Agric. Food Chem., 21, 923 /1973/
4. Murray, K.E. and Stanley, G., J. Chromatog., 34, 174 /1968/
5. Hajós, A., Complex hydrides, Akadémiai Kiadó, Budapest, 1973, p. 46
6. Hanna, J.G., Chemical and physical methods of analysis -in: Patai, S.: The chemistry of the carbonyl group, Interscience Publishers, London-New York-Sydney, 1966, p. 376

A GAS CHROMATOGRAPHIC POLARITY PARAMETER AND ITS APPLICATION IN STUDIES OF QUANTITATIVE STRUCTURE-OLFACTORY ACTIVITY RELATIONSHIPS IN PHENOLS

ROMAN KALISZAN, HENRYK LAMPARCZYK, MARIUSZ PANKOWSKI, BARBARA DAMASIEWICZ, ANTONI NASAL and JANUSZ GRZYBOWSKI

Faculty of Pharmacy, Medical Academy, 80-416 Gdańsk, Poland

SUMMARY

A new chemical descriptor, chromatographic polarity parameter, intended for use in QSAR calculations has been derived.
The descriptor is defined by the formula $I - b\,MR$, where I is
the Kováts retention index determined by gas-liquid chromatography and MR is the molar refraction of the substituent. In
order to calculate b, a measure of the ability of a stationary
phase to form dispersive bonds, the retention data are needed
for the compounds studied which are obtained on two stationary
phases of different polarities. The proposed polarity parameter
satisfactorily correlates with the dipole moments calculated
quantum-chemically. For a group of 19 substituted phenols the
olfactory thresholds have been determined in experiments with
8-10 human subjects. Significant correlations have been obtained
in describing the olfactory activity as a square function of
the hydrophobicity parameter corrected for ionization. The
chromatographic polarity parameter has been proved to be a convenient and easily determined correction for ionization for the
hydrophobicity parameter. It may also be of value for studies
where polar drug-receptor interactions are expected.

INTRODUCTION

Since the publications of Iwasa et al. (1) and Boyce and
Milborrow (2) in 1965 chromatography has become a powerful tool

for the determination of molecular structural data employed in studies of quantitative structure-activity relationships (QSAR) (3,4).

The capacity of the dipole-dipole interaction of a drug with the biological receptor is often assumed to be of import- ance for its bioactivity. Since the determination of a measure of molecular polarity, such as e.g., the dipole moment, are rather tedious there is a need to develop a convenient method for the evaluation of a quantity related to polarity. Such an attempt was undertaken earlier (5) and is now further developed here employing gas chromatographic retention data determined on stationary phases of different polarity. The application of the proposed measure of polarity to QSAR is discussed based on the olfactory activity of a group of substituted phenols previously determined and expressed as the detection thresholds (6).

THEORETICAL

It was observed earlier (7) that gas-liquid chromato- graphic retention indices of fatty acid methyl esters on SE-30 and SILAR 5CP stationary phases were related by the empirical equation

$$I = K_1 \cdot P + K_2 \cdot + K_3 \tag{1}$$

where P is a parameter related to the molecular polarity of the solutes, is the connectivity index (8, 9) characterizing the ability of the solutes to undergo dispersive interactions with the stationary phases and $K_1 - K_3$ are constants. Similar equa- tion, with the square of the dipole moment instead of P, was proposed by Gassiot-Matas and Firpo-Pamies (10). The connec- tivity index, X, is easily calculated for hydrocarbons and organic compounds containing nitrogen and oxygen. However, in the case of other atoms, e.g., halogens, X values are uncertain and generally the molecular refractivity, MR, seems to be the more reliable parameter for the description of the ability to form dispersive bonds. Thus, our starting relationship has the form

$$I = a \cdot P + b \cdot MR + c \tag{2}$$

where $\underline{a}$ can be considered as a measure of the stationary phase polarity and $\underline{b}$ as a measure of its ability to undergo dispersive interactions with a solute. Rearranging eqn. 2 one obtains

$$I - b \cdot MR = a \cdot P + c \tag{3}$$

Thus, if eqn. 2 holds, the quantity $\underline{I} - \underline{b \cdot MR}$ should be linerarly related to the polarity of the solute. The molecular refractivity, $\underline{MR}$, can easily be calculated (11, 12). Thus, if the retention indices can be determined and the value of $\underline{b}$ is known, we have a convenient, quantitative measure for the polarity of the solutes.

The constant $\underline{b}$ is calculated as follows. If the retention indices for any given compound on two phases of different polarity are available then one can write that

$$I_P = a_1 \cdot P + b_1 \cdot MR + c_1 \tag{4}$$

$$I_{NP} = a_2 \cdot P + b_2 \cdot MR + c_2 \tag{5}$$

where $\underline{I_P}$ and $\underline{I_{NP}}$ are retention indices on polar and non-polar phases, respectively. Taking $\underline{P}$ from eqn. 5

$$P = (I_{NP} - b_2 \cdot MR - c_2)/a_2 \tag{6}$$

eqn. 4 can be rewritten as:

$$I_P = a_1 \cdot (I_{NP} - b_2 \cdot MR - c_2)/a_2 + b_1 \cdot MR + c_1 \tag{7}$$

After rearrangement one obtains

$$I_P = \frac{a_1}{a_2} \cdot I_{NP} - \left(\frac{a_1}{a_2} \cdot b_2 - b_1 \right) \cdot MR - \frac{a_1}{a_2} \cdot c_2 + c_1 \tag{8}$$

or, introducing the new constants $k_1 = \dfrac{a_1}{a_2}$, $k_2 = \dfrac{a_1}{a_2} \cdot b_2 - b_1$

and $k_3 = -\dfrac{a_1}{a_2} \cdot c_2 + c_1$

$$I_P = k_1 \cdot I_{NP} - k_2 \cdot MR + k_3 \qquad (9)$$

The constants $\underline{k}_1 - \underline{k}_3$ can be determined and statistically evaluated from the regression analysis of the experimental by determined retention indices and molecular refractivity data for a sufficiently large set of compounds. A statistically significant relationship of the type of eqn. 9 proves the validity of the assumptions expressed by eqn. 2. Knowing $\underline{k}_1$ and $\underline{k}_2$ one can then calculate $\underline{b}$.

Let us first consider the situation when $b_1 = b_2$, i.e., the two phases, polar and non-polar, are of equal ability to form dispersive bonds with a solute. To a first approximation one can assume this to be the case when the two phases have similar molecular weights. Then:

$$b_1 = b_2 = b = k_2/(k_1 - 1) \qquad (10)$$

After calculation of $\underline{b}$ one can use eqn. 3 to determine the chromatographic polarity parameter, $(\underline{I} - \underline{b} \cdot \underline{MR})$.

A more complex situation exists if the abilities of the phases to form dispersive bonds, $\underline{b}_1$ and $\underline{b}_2$, significantly differ. Then, the value for one of the two phases must be known and the second value can be expressed as

$$b_1 = k_1 \cdot b_2 - k_2 \quad \text{or} \quad b_2 = (k_2 + b_1)/k_1 \qquad (11)$$

A possible way to obtain data related to the ability of a chosen standard phase to form dispersive bonds would be to find a relationship between the molecular refractivity and the retention indices for a group of non-polar compounds. The phase chosen should be as non-polar as possible.

682

Alternatively different values of b calculated from eqn. 11 can be compared with those obtained from eqn. 2. In order to get a statistically significant relationship for an equation of the type of eqn. 2 a certain number of retention indices, polarity measures and molecular refractivities are required. If dipole moments can be assumed as the polarity measures then substituted phenols offer a group of compounds for which quite a number of such data are available in the literature.

EXPERIMENTAL

Chromatography

The Kováts retention indices for 43 phenols on dimethyl-polysiloxane (SE-30), 3-cyanopropylmethylpolysiloxane (OV-225) and polyneopentyl glycol adipate (NGA) coated on Chromosorb W HMDS (80-100 mesh) have been reported earlier (13). The measurements were made at $150^{o}C$. The numerical values are given in Table I along with the molar refractivities calculated as a sum of fragmental refractivities according to Hansch et al. (11).

For 20 of the phenols studied dipole moment values determined experimentally in benzene solutions at $25^{o}C$ (or at $20^{o}C$ in four cases) have been found in the literature (14-20).

In Table I the quantum-chemically calculated dipole moments (5) are also given for 43 phenols. The CNDO/2-MO method was used for the calculations. In cases where more than one energetically favoured conformer exists, the value μ_{calc} represents the arithmetic mean of the μ values of the corresponding conformers.

Determination of the Relative Olfactory Thresholds

A group of 19 phenols was chosen for the biological experiment. The compounds selected are intensely odourous at low concentrations and their odour is unambiguously classified as phenolic by persons having some chemical experience.

A series of solutions of a given compound in distilled water was placed in open glass plates. A group of 8-10 male and

Table I. Retention indices, molar refractivities, moments and chromatographic polarity

No.	Phenol	Kováts retention indices		
		SE-30	OV-225	NGA
1	2		3	
1	2-CH_3	1035	1587	1742
2	4-CH_3	1059	1654	1813
3	3-CH_3	1065	1648	1782
4	2,6-/$CH_3/_2$	1098	1593	1716
5	2,4-/$CH_3/_2$	1134	1660	1825
6	3-C_2H_5	1160	1742	1898
7	4-C_2H_5	1162	1746	1890
8	3,5-/$CH_3/_2$	1163	1706	1877
9	2,3-/$CH_3/_2$	1169	1693	1857
10	2,4-Cl_2	1183	1708	1877
11	4-Cl	1192	1922	2058
12	3-Cl	1194	1911	2061
13	2,4,6-/$CH_3/_3$	1204	1621	1778
14	2,6-Cl_2	1206	1727	1871
15	4-OCH_3	1210	1930	2050
16	3-OCH_3	1211	1940	2083
17	2,3,5-/$CH_3/_3$	1260	1823	1960
18	4-Br	1274	2054	2191
19	3-CH_3-4-Cl	1283	2025	2135
20	4-NH_2	1314	2154	2277
21	4-OH	1334	2330	2515
22	3-NH_2	1335	2219	2352
23	2,4,6-Cl_3	1349	1928	2067
24	2,4,5-Cl_3	1362	2039	2158
25	3-OH	1368	2371	2576
26	3,5-Cl_2	1391	2217	2343
27	4-J	1398	2230	2348
28	4-CO_2CH_3	1500	2376	2461
29	4-$COCH_3$	1578	2478	2429
30	2-NH_2	1242	2039	2196

experimentally measured and calculated dipole
parameters for a group of substituted phenols

Molar refractivity MR	Dipole moment values (benzene 25°C)		Chromatographic polarity parameters I_{NGA} – 21.3825 MR
	μ_{obs}	μ_{calc}*	
4	5		6
32.83	1.45	1.72	1040
32.83	1.61	1.87	1111
32.83	1.58	1.79	1080
37.45	1.38**	1.69	915
37.45	1.39**	1.77	1024
37.48	–	1.60	1096
37.48	–	1.85	1088
37.45	1.55**	1.82	1076
37.45	1.25**	1.74	1056
38.21	1.59	2.12	1060
33.21	2.19	2.36	1348
33.21	2.14	2.03	1351
42.07	1.40	1.78	878
38.21	–	3.20	1054
35.05	1.92	2.21	1300
35.05	–	2.30	1334
42.07	–	1.78	1060
36.06	2.19	2.70	1420
37.84	–	2.02	1326
32.60	–	2.50	1580
28.03	1.40	1.61 (3.21)	1916
32.60	1.83	2.53 (2.61)	1655
43.21	1.62	1.67	1143
43.21	–	1.99	1234
28.03	2.07	3.49	1977
38.21	2.18	2.17	1526
41.02	2.13	–	1471
40.05	–	2.71	1604
38.36	–	3.15	1709
32.60	–	2.85	1499

Table I

1	2		3	
31	3-Br	1270	2069	2214
32	2-iso-C_3H_7-5-CH_3	1271	1776	1932
33	2,6-/tert.-$C_4H_9/_2$-4-CH_3	1494	1782	1830
34	2-OCH_3	1095	1544	1627
35	2-NO_2	1149	1556	1703
36	2,6-/$OCH_3/_2$	1347	1936	2014
37	2-OCH_3-4-C_3H_7	1392	1810	1884
38	2-OCH_3-4-CHO	1447	2199	2235
39	2,6-/$OCH_3/_2$-4-CH_3	1473	2076	2106
40	2-OCH_3-4-$COCH_3$	1531	2283	2326
41	2,6-/$OCH_3/_2$-4-C_3H_7	1624	2254	2256
42	2,6-/$OCH_3/_2$-4-$COCH_3$	1849	2685	2683
43	2-OCH_3-4-CH_2-CH=CH_2	1367	1848	1923

*Arithmetic mean of the μ values for the energetically
favourable conformers.

**Determined in benzene at $20^\circ C$.

female volunteers took a smell of the samples close to their
surface starting from the most diluted sample. Four persons of
the group were taking part in each experiment while the other
4-6 persons were changed over. Parallel with a given derivative
an experiment was carried on with unsubstituted phenol as the
standard. All the determinations were carried out in the same
airconditioned room at a temperature $21 \pm 1^\circ C$. As a result of
these studies, the relative measure of bioactivity log A has
been defined as equal to log C_p/C_x, where C_p and C_x are the
lowest detected molar concentrations of phenol and the derivative studied, respectively. The data are given in Table II as
average values for a group of 8-10 subjects studied.

In Table II the hydrophobicity parameter, Π, is also given
obtained by the subtraction of the partition coefficient of
phenol in n-octanol-water from that of the given derivative
(21-23).

(cont.)

4	5	6	
36.06	–	2.34	1443
45.75	–	1.82	954
70.01	–	1.75	333
35.05	–	2.12	878
34.54	3.10	3.82	964
41.89	–	3.00	1118
48.98	–	2.28	837
40.90	–	3.39	1360
46.51	–	3.07	1112
38.36	–	2.89	1506
55.82	–	3.20	1062
52.04	–	4.49	1570
48.51	–	2.18	886

Also, the hydrophobicity parameter corrected for ionization, $(\Pi + C_D)$, has been considered. The value of C_D was calculated according to the formula:

$$C_D = \log \frac{K_{a_x} - \left[H^+\right]}{K_{a_p} - \left[H^+\right]} \tag{12}$$

where $\left[H^+\right]$ is the hydrogen ion concentration at the physiological pH = 7.4 and K_{a_x} and K_{a_p} are the dissociation constants for a derivative and phenol, respectively. K_a or pK_a values were found in the literature (22, 24) or determined spectrophotometrically. Hammett's σ values were taken from the compilation by Hansch et al. (11).

Table II. Biological and physicochemical

No.	Phenol	Activity log A	$\Pi = \log P_x - \log P_p$	Π^2	Π corrected for ionization $(\Pi + C_D)$
1	H	-0.04	0	0	0
2	2-Cl	2.12	0.70	0.49	0.70
3	4-Cl	2.01	0.92	0.85	0.92
4	3-Br	2.06	1.16	1.35	1.15
5	4-Br	1.76	1.15	1.32	1.15
6	$2,4,5-Cl_3$	1.36	2.86	8.18	2.32
7	$4-NO_2$	0.32	0.44	0.19	0.0
8	$3-CH_3$	1.45	0.50	0.25	0.50
9	$4-CH_3$	1.62	0.46	0.21	0.46
10	$2,4-/CH_3/_2$	1.36	0.97	0.94	0.97
11	$2,5-/CH_3/_2$	1.60	1.05	1.10	1.05
12	$2,6-/CH_3/_2$	1.69	0.89	0.79	0.89
13	$3-CH_3-4-Cl$	0.95	1.49	2.22	1.49
14	$3-C_2H_5$	1.65	0.93	0.86	0.93
15	$4-OCH_3$	-0.30	-0.13	0.02	-0.13
16	$2,6-/OCH_3/$	0.18	-0.26	0.07	-0.26
17	$2-OCH_3-4-COCH_3$	-0.18	-0.24	0.06	-0.25
18	$4-COCH_3$	-0.64	-0.11	0.01	-0.24
19	$2-iso-C_3H_7-5-CH_3$	1.95	1.87	3.58	1.87

RESULTS AND DISCUSSION

The relationship between the retention indices on OV-225 and SE-30 has a form analogous to eqn. 9

$$I_{OV-225} = 1.95 \, (\pm 0.16) \cdot I_{SE-30} - 22.08 \, (\pm 3.56) \cdot MR + 285.76 \tag{13}$$

$$n = 43 \, , \; r = 0.9691 \, , \; s = 71$$

data for a series of phenol derivatives

$/\Pi + C_D/^2$	pK_a	$\Sigma\sigma$ Hammett's constant for substituents	Retention index on NGA phase I_{NGA}	Chromatographic polarity measure $I_{NGA} - 21.3825\ MR$
0	10.01	0	1703	1100
0.49	9.52	0.23	1555	845
0.85	9.36	0.33	2058	1348
1.33	9.11	0.39	2214	1443
1.32	9.33	0.23	2191	1420
5.37	7.00	0.83	2158	1234
0.0	7.16	0.78	2761	2022
0.25	10.09	-0.07	1782	1080
0.21	10.26	-0.17	1813	1111
0.94	10.47	-0.34	1825	1024
1.10	10.32	-0.24	1770	969
0.79	10.60	-0.34	1716	915
2.22	9.4	0.16	2135	1326
0.86	10.09	-0.07	1898	1088
0.02	10.21	-0.27	2050	1300
0.07	11.30	-0.54	2014	1118
0.06	9.10	0.23	2326	1506
0.06	7.88	0.50	2529	1709
3.58	10.49	-0.22	1932	954

where n is the number of compounds considered, r is the multiple correlation coefficient and s is the standard deviation from the regression equation. The numbers in parentheses represent 95% confidence limits.

The corresponding equation in the case of the phases NGA and SE-30 has the form:

$$I_{NGA} = 1.81 \; (\pm 0.20) \cdot I_{SE-30} - 24.50 \; (\pm 4.36) \cdot MR +$$
$$+ 682.59 \tag{14}$$
$$n = 43, \; r = 0.9472 \; , \; s = 87$$

To calculate b_1 by eqn. 11 the b_2 value must be known apart of the values of k_1 and k_2 obtained from eqn. 14.

To obtain b we tried to use eqn. 2 assuming the dipole moment of the phenols as the polarity parameter, P. We have found experimental dipole moments for a group of 20 compounds. For four compounds, dipole moments determined in benzene at ambient conditions did not fit eqn. 2. These are 4-OH, 3-OH and 3-NH$_2$ substituted phenols for which the experimental dipole moments are significantly lower than expected from eqn. 2, and 2-nitrophenol, the experimental dipole moment of which is too high to fit eqn. 2. The dipole moments of the remaining 16 compounds under chromatographic conditions are (to a first approximation) linearly related to the experimentally measured values. Considering all the limitations, the following equations can be written which support the hypothesis:

$$I_{SE-30} = 66.67 \; (\pm 14.42) \cdot \mu_{obs}^2 + 25.34 \; (\pm 5.22) \cdot MR +$$
$$+ 57.53 \tag{15}$$
$$n = 16 \; , \; r = 0.9679 \; , \; s = 31$$

$$I_{OV-225} = 164.72 \; (\pm 36.61) \cdot \mu_{obs}^2 + 24.48 \; (\pm 12.90) \cdot MR +$$
$$+ 408.33 \tag{16}$$
$$n = 16 \; , \; r = 0.9472 \; , \; s = 77$$

$$I_{NGA} = 155.69 \; (\pm 33.63) \cdot \mu_{obs}^2 + 23.99 \; (\pm 12.19) \cdot MR +$$
$$+ 599.81 \tag{17}$$
$$n = 16 \; , \; r = 0.9477 \; , \; s = 73$$

The two-parameter eqns. 15-17 are statistically highly significant. The correlation coefficients for the corresponding one-parameter equations relating the retention indices to μ_{obs}^2 and MR are much lower or of no statistical value (Table III). In the next step of our investigations, we compared the b values obtained from eqns. 15-17 with those calculated from the

Table III. Correlation coefficients, R, and standard deviations, s, for the linear equations $I = \alpha X + \beta$ relating retention index, I, to dipole moments or molar refractivities, X, for a group of phenols

I	X	R	s
I_{SE-30}	μ^2_{obs}	0.6065	95
I_{OV-225}	μ^2_{obs}	0.8692	114
I_{NGA}	μ^2_{obs}	0.8641	110
I_{SE-30}	MR	0.6479	91
I_{OV-225}	MR	0.2330	225
I_{NGA}	MR	0.2463	213

coefficients in eqns. 13 and 14. For the calculation of b_1 by eqn. 11, the value $b_2 = 25.3412$ has been assumed as obtained from eqn. 15. Then, using eqn. 13. b was calculated for the OV-225 phase as 27.3510. The corresponding value for the NGA phase calculated from eqn. 14 is 21.3825.

As can be concluded from the reports by Karger et al. (25) and R.P.W. Scott (26) the polar term in eqn. 2-type relationships reflects more than just the charge distribution in the molecule. It also includes a steric component and reflects the ability to form hydrogen bonds between the solutes and the stationary phases. Therefore, a chromatographic polarity parameter might not be ideally correlated to the dipole moment but, on the other hand, it might be able to simulate the capacity of a drug molecule to approach its binding site in a living system.

The chromatographic polarity parameter, being dynamic in nature, rather reflects the actual dipole moment of the solute under chromatographic conditions. On the other hand, calculated dipole moments represent the molecular conformations in the gas phase which may well differ from those in the adsorbed state.

In such a situation the relationship between the chromatographic polarity parameter and the quantum-chemically calculated
dipole moments

$$I_{NGA} - 21.3825 \cdot MR = 98.85 \; \mu_{calc}^2 + 810.62 \tag{18}$$
$$n = 27, \; r = 0.919, \; s = 8.47$$

is quite satisfactory. Eqn. 18 was obtained after elimination
of all phenols with space-consuming substituents (Nos. 33,37,
41,43) as well as 2,6-disubstitution (Nos. 4,13,14,23,33,36,39,
41,42), also No. 34 (steric hindrance and hydrogen bonding).
No. 35 (hydrogen bonding) and No. 38 (thermal reaction ?),
leaving 27 phenols. This procedure seems to be reasonable because in the remaining 27 phenols the interaction between the
phenol and the stationary phase is dominated by the electronic
properties of the molecule (dipole moments), whereas in all the
excluded cases additional factors play a significant role.

Finally, it is interesting to investigate the usefulness
of the proposed chromatographic polarity parameter for QSAR.

At first we studied the relationship between this parameter
and the other measures of electronic properties of a group of
phenols for which biological data have been collected (Table
II). The following equations were obtained:

$$I_{NGA} - 21.3825 \cdot MR = -186.4 \cdot pK_a + 3020 \tag{19}$$
$$n = 19, \; r = 0.73, \; s = 215$$

$$I_{NGA} - 21.3825 \cdot MR = 528.6 \; \Sigma\sigma + 1198 \tag{20}$$
$$n = 19, \; r = 0.69, \; s = 227$$

The correlations, although significant at 99% significance
level are not very high. Probably pK_a and $\Sigma\sigma$ are not very
precise measures of the ability of a molecule to undergo polar
interactions with chromatographic stationary phases.

The relationship between the olfactory activity of the
phenols studied and the hydrophobicity parameter, uncorrected
for ionization, is as follows:

$$\log A = -0.59 \ (\overset{+}{-} 0.28) \cdot \Pi^2 + 2.03 \ (\overset{+}{-} 0.70) \cdot \Pi +$$
$$+ 0.24 \ (\overset{+}{-} 0.37) \tag{21}$$
$$n = 19 \ , \ r = 0.864 \ , \ s = 0.50$$

An analogous relationship with corrected hydrophobicity parameter has the form of eqn. 22:

$$\log A = -0.80 \ (\overset{+}{-} 0.33) \cdot (\Pi + C_D)^2 +$$
$$+ 2.23 \ (\overset{+}{-} 0.62) \cdot (\Pi + C_D) + 0.33 \ (\overset{+}{-} 0.29) \tag{22}$$
$$n = 19 \ , \ r = 0.909 \ , \ s = 0.41$$

Introduction of the chromatographic polarity parameter into
eqn. 22 gave no significant improvement of the correlation.
However, the introduction of $I_{NGA} - 21.3825 \cdot MR$ into eqn. 21
gave a correlation similar to that in eqn. 22, where hydrophobicity corrected for ionization was applied:

$$\log A = -0.54 \ (\overset{+}{-} 0.25) \cdot \Pi^2 + 1.85 \ (\overset{+}{-} 0.63) \cdot \Pi -$$
$$- 0.0009 \ (\overset{+}{-} 0.0008) \cdot (I_{NGA} - 21.3825 \cdot MR) +$$
$$+ 1.40 \ (\overset{+}{-} 1.07) \tag{23}$$
$$n = 19 \ , \ r = 0.907 \ , \ s = 0.43$$

As evident from eqn. 23 the chromatographic polarity parameter plays here the role of an approximate correction for
ionization for the hydrophobicity parameter. It is assumed,
however, that the chromatographic parameter could also be of
value for studies where polar drug-receptor interactions were
expected. In fact, in the case of phenols, a number of classical
electronic data are available. An opposite situation is
observed for the majority of non-aromatic drug systems. As the
chromatographic polarity parameter can easily be determined
also in the case of compounds for which quantification of
electronic properties is troublesome, it may be considered in
studies where covalent or ionic drug-receptor bonds are likely
to be formed.

REFERENCES

1. J. Iwasa, T. Fujita and C. Hansch, J. Med. Chem., <u>8</u>, 150 (1965)

2. C.B.C. Boyce and B.V. Milborrow, Nature (London), <u>208</u> 537 (1965)

3. E. Tomlinson, J. Chromatogr., <u>113</u> 1 (1975)

4. R. Kaliszan, J. Chromatogr., <u>220</u> 71 (1981)

5. R. Kaliszan and H.-D. Höltje, J. Chromatogr., <u>234</u> 303 (1982)

6. R. Kaliszan, M. Pankowski, L. Szymula, H. Lamparczyk, A. Nasal, B. Tomaszewska and J. Grzybowski, Pharmazie, <u>37</u> 499 (1982)

7. R. Kaliszan, Chromatographia, <u>12</u> 171 (1979)

8. M. Randić, J. Chromatogr., <u>161</u> 1 (1978)

9. L.B. Kier and L.H. Hall, Molecular Connectivity in Chemistry and Drug Research, Academic Press, New York, 1976.

10. M. Gassiot-Matas and G. Firpo-Pamies, J. Chromatogr., <u>187</u> 1 (1980)

11. C. Hansch, A. Lec, S.H. Unger, K.H. Kim, D. Nikaitani and E.J. Lien, J. Med. Chem., <u>16</u> <u>1207</u> (1973)

12. A.J. Vogel, W.T. Cresswell and J. Leicester, J. Phys. Chem., <u>58</u> 174 (1954)

13. J. Grzybowski, H. Lamparczyk, A. Nasal and A. Radecki, J. Chromatogr., <u>196</u> 217 (1980)

14. H.L. Doule, Z. Phys. Chem. (Leipzig), <u>18</u> 146 (1932)

15. R. Perrin and P. Issortel, Bull. Soc. Chim. Fr., 1083 (1967)

16. Ch. Sun and Ch. Lin. J. Chin. Chem. Soc., <u>5</u> 39, (1937)

17. A. Koll, H. Ratajczak and L. Sobczyk, Rocz. Chem., 44 825 (1970)

18. J.J. Lander and W.J. Svirbely, J. Am. Chem. Soc., <u>67</u> 322 (1945)

19. O. Hassel and E. Naeshagen, Z. Phys. Chem. (Leipzig), <u>12</u> 79 (1931)

20. J.W. Williams and J.M. Fogelberg, J. Am. Chem. Soc. <u>52</u>, 1365 (1930)

21. T. Fujita, J. Med. Chem., <u>9</u> 797 (1966)

22. T. Fujita, J. Iwasa and C. Hansch, J. Am. Chem. Soc., <u>86</u>, 5175 (1964)

23. R.F. Rekker, The Hydrophobic Fragmental Constant, Elsevier, Amsterdam, 1977.

24. Landolt-Börnstein, Zahlenwerke und Funktionen, Springer, Berlin, 1960.

25. B.L. Karger, L.R. Snyder and C. Eon, J. Chromatogr., <u>125</u>, 71 (1976)

26. R.P.W. Scott, J. Chromatogr., <u>122</u> 35 (1976)

Chromatography, the State of the Art
H. Kalász and L.S. Ettre (Eds)

COMPARISON OF THE POROUS POLYMERS CHROMOSORB 101 AND CHROMOSORB 102 AS SUPPORTS COATED WITH ETHOFAT 60/25 IN GAS CHROMATOGRAPHY

D. BARCELÓ, M.T. GALCERÁN and L. EEK*

Department of Analytical Chemistry, Faculty of Chemistry,
University of Barcelona, Diagonal 647, Barcelona-28, Spain
*Derivados Forestales, S.A., Pº Sant Joan 15. Barcelona-10.
Department of Inorganic and Analytical Chemistry,
E.T.S.E.I.B. Polytechnical University of Barcelona,
Diagonal 647, Barcelona-28, Spain

SUMMARY

In the present paper we compare the behaviour of Chromosorb 101 and Chromosorb 102 porous polymers as supports coated with Ethofat 60/25 using a series of compounds of varying polarity.

We calculated the adsorption constants for both supports and the surface areas for Chromosorb 102. These surface areas were related to the molar volumes of each solute and the solubility parameters of the compounds, adsorbent and liquid phase.

We have compared the scanning electron microphotographs of both supports coated with Ethofat 60/25 observing a similar appearance and showing little holes in the surface which can be responsible for the observed adsorption.

INTRODUCTION

In three previous papers (1,2,3) we have studied the solute-stationary phase-adsorbent interaction phenomena associated with the use of active Chromosorb 101 support in gas chromatography.

In the present paper we complete this study comparing the behaviour of the previously used Chromosorb 101 with Chromosorb 102 porous polymer support, both coated with the same liquid phase, Ethofat 60/25.

Both supports are styrene-divinylbenzene polymers having different surface areas and different average pore diameters.

In order to study the unlike behaviour, we calculate the adsorption terms of some solutes of varying polarity on both supports. These terms are greater for Chromosorb 102 than for Chromosorb 101, showing a higher effect of the former on the chromatographic process. We also calculate the available surface area of Chromosorb 102 support, assuming that the surface area of Chromosorb 101 is 50 m^2/g, and that the adsorption constant of Chromosorb 102 should be the same as that of Chromosorb 101. This available surface area of Chromosorb 102 support has a different value according to the studied compound, but we have considered an average surface area of 95 m^2/g, which represents 30% of the surface area indicated by the manufacturer.

The observed differences in the calculated surface area values can be attributed to the unlike molar volumes of the solutes studied and to the interactions between these solutes, the adsorbent and the small amount of stationary phase coating the support in the interval we calculated the adsorption terms. These interactions have been related to the specific solubility parameters of the solute, adsorbent and liquid phase, obtaining relatively constant relations between these variables and the calculated surface areas.

Finally, we have compared the scanning electron microphotographs of both supports coated with Ethofat 60/25 and we have noticed a different appearance according to the pore diameter of each support using an ordinary SEM apparatus with a resolution of 200 Å, but when increasing the magnification and the resolution of the instrument to 60 Å, we have observed a similar appearance of both supports suggesting that the retention mechanism is similar.

EXPERIMENTAL

We chose 25 compounds of varying polarity (see Tables I and II). The columns used were 2 m x 1/8 in. packed with Chromosorb 102 80/100 coated with different amounts of Ethofat 60/25 (a monostearate of polyoxyethylene with an average molecular

Table I. Net retention volume V_N (cm^3) of various substances on Chromosorb 101 coated with Ethofat 60/25.

COLUMNS	1	2	3	4	5	6	7	8	9	10	11
% ETHOFAT 60/25	0	0.4	1.1	2.6	5.5	10	15	20	25	30	40
$V_N / W_S \cdot 10^2$	0	0.4	1.0	2.9	5.8	11.1	17.6	25	33	43	66.7
n-PENTANE	90	86	55	42	22	19	19	19	19	19	19
n-HEXANE	300	288	240	139	96	44	43	43	41	40	36
n-HEPTANE	900	756	696	439	223	96	91	91	85	80	73
n-OCTANE	--	--	--	--	427	200	183	183	167	156	150
METHANOL	35	29	27	28	29	30	36	42	48	54	67
ETHANOL	60	58	55	50	54	54	55	65	73	84	102
ISOPROPANOL	100	91	84	74	76	79	80	82	90	98	114
TERT-BUTANOL	125	120	103	92	80	82	84	86	90	96	112
n-PROPANOL	180	158	132	122	123	124	128	136	149	168	200
DIETHYL ETHER	113	98	91	80	69	35	34	34	33	33	33
DI-ISOPROPYL ETHER	420	396	372	160	133	53	53	51	48	48	49
METHYL FORMATE	47	41	39	39	37	29	30	31	34	40	43
ETHYL FORMATE	125	120	103	92	86	50	55	60	63	70	80
METHYL ACETATE	144	139	132	106	96	64	64	69	74	80	95
ETHYL ACETATE	355	317	271	224	205	109	106	116	122	133	154
ISOPROPYL ACETATE	667	576	480	415	301	135	137	145	149	155	180
METHYL PROPIONATE	399	367	336	252	240	125	131	139	144	149	166
ETHYL PROPIONATE	--	--	--	--	--	226	200	235	236	240	252
FORMALDEHYDE	30	34	38	45	57	92	151	234	293	294	300
ACETALDEHYDE	35	29	27	27	27	27	28	29	30	31	34
PROPIONALDEHYDE	114	108	102	72	71	56	56	57	60	70	71
ACETONE	133	125	115	66	60	50	56	61	64	71	82
METHYL ETHYL KETONE	312	264	245	219	181	111	119	126	135	146	162
METHYLAL	132	120	108	92	79	45	48	50	52	56	61
WATER	14	17	22	24	28	32	45	59	71	82	101

Table II. Net retention volume V_N (cm^3) of various substances on Chromosorb 102 coated with Ethofat 60/25.

COLUMNS	12	13	14	15	16	17	18	19	20
% ETHOFAT 60/25	0	0.3	1	4.2	13.1	18.8	23	30	40
$V_N / W_S \cdot 10^2$	0	0.3	1	4.5	15.2	23.1	29.9	41.7	66.7
n-PENTANE	312	262	216	65	60	50	43	41	40
n-HEXANE	840	760	516	169	134	110	94	90	84
n-HEPTANE	--	--	--	--	293	226	192	187	175
n-OCTANE	--	--	--	--	631	478	372	360	348
METHANOL	46	38	36	31	34	43	52	65	72
ETHANOL	108	96	86	67	62	72	85	103	136
ISOPROPANOL	228	190	161	110	89	98	107	120	144
TERT-BUTANOL	264	241	211	113	110	115	118	127	153
n-PROPANOL	312	305	256	169	158	180	199	223	264
DIETHYL ETHER	288	269	218	86	70	60	55	55	56
DI-ISOPROPYL ETHER	--	1032	720	336	149	118	106	102	100
METHYL FORMATE	96	79	72	50	46	41	48	49	52
ETHYL FORMATE	264	241	211	113	94	86	94	96	98
METHYL ACETATE	300	245	216	118	103	96	106	108	112
ETHYL ACETATE	720	672	504	326	199	180	178	192	216
ISOPROPYL ACETATE	--	--	--	--	295	240	223	232	252
METHYL PROPIONATE	--	--	600	367	235	201	211	222	235
ETHYL PROPIONATE	--	--	--	--	427	374	367	376	400
FORMALDEHYDE	48	45	43	41	91	118	151	180	384
ACETALDEHYDE	62	53	48	36	31	31	35	37	42
PROPIONALDEHYDE	240	194	175	112	86	79	70	74	88
ACETONE	211	178	158	108	83	77	80	85	95
METHYL ETHYL KETONE	648	554	432	300	192	180	176	187	204
METHYLAL	300	262	206	118	86	83	83	83	84
WATER	19	19	19	22	58	62	74	91	114

Table III Typical physical properties of the two supports

	Chromosorb 101	Chromosorb 102
Type	STY-DVB	STY-DVB
Free fall density (g/ml)	0.30	0.29
Surface area (m^2/g)	50	300-400
Average pore diameter (μm)	0.3-0.4	0.0085
Water affinity	hydrophobic	hydrophobic
Color	white	white
Temperature limit (isothermal) ^{o}C	275	250

mass of 938) as the stationary phase. Table III lists the physical properties of both supports.

The gas chromatograph used was a Perkin-Elmer Model 990 equipped with a thermal conductivity detector. The experimental conditions were as indicated previously (1).

For the study by scanning electron microscope, the column packings of Ethofat on Chromosorb 101 and Chromosorb 102 were mounted on specimen holders coated with a 400-600 Å layer of gold in a vacuum evaporator fitted with a sputtering diode. They were examined with a Stereoscan Model S-4 scanning electron microscope (Cambridge Instruments), using an accelerator voltage between 5 and 10 kV and with a magnification of 5000 and 10,000. The samples of Ethofat on Chromosorb 102 were examined with a JEOL JSM-35C scanning microscope with a resolution of 60 Å using an accelerator voltage of 25 kV and with a magnification of 20,000 and 48,000.

RESULTS AND DISCUSSION

Tables I and II list the different values of V_N of the compounds studied on the columns used while Figs 1 and 2 present the plots of V_N/W_S vs. W_L/W_S for some of the solutes with varying polarity [V_N = net retention volume, W_S = weight of support in the column, W_L = weight of liquid phase in the column].

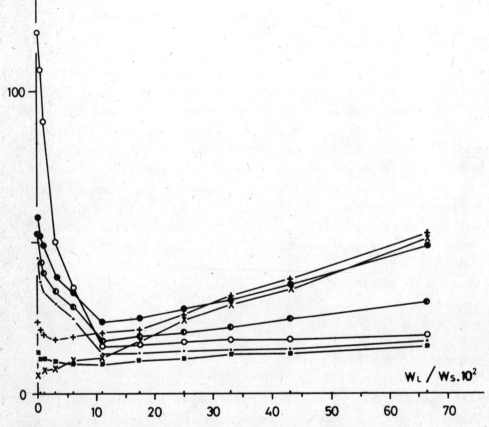

Fig. 1. Relationship between V_N/W_S and liquid loading, $W_L/W_S \cdot 10^2$. Columns: Ethofat 60/25 on Chromosorb 101
(o) n-Hexane, (+) Ethanol, (.) Diethyl ether, (●) Methyl acetate, (■) Acetaldehyde, (◐) Methylal and (x) Water.

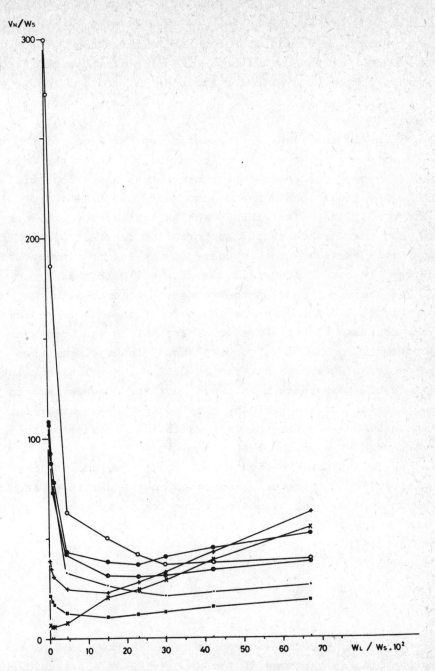

Fig. 2. Relationship between V_N/W_S and liquid loading, $W_L/W_S \cdot 10^2$. Columns: Ethofat 60/25 on Chromosorb 102 (o) n-Hexane, (+) Ethanol, (.) Diethyl ether, (●) Methyl acetate, (■) Acetaldehyde, (◉) Methylal and (x) Water.

We can see that the graphs obtained with the two different supports are very similar for the same compound, and the retention volumes increase with the amount of stationary phase for the polar compounds, while for the nonpolar and intermediate polar compounds they decrease at low percentages, increasing afterwards.

It can be observed that the different sections of the graphs have different widths in both supports. For the Chromosorb 101 support there is a decrease with a high slope till 2.6% of the liquid phase and a minor decrease till 10% of liquid phase, then after this point an increase. On the other hand, for the Chromosorb 102 support, the first section of the graphs reaches 4.2% and the second one is larger since straight sections are not noticed till 23% of the liquid phase.

The sharp decrease on the retention volumes with the amount of liquid phase in the low percentage zone is explained by the decrease in the superficial area of the support when it is covered. Chromosorb 102 support shows higher slopes than Chromosorb 101 support that can be due to the relatively larger decrease of its surface area. Such decrease depends in turn on the number of pores covered by the liquid phase, larger for the Chromosorb 102 than for the Chromosorb 101, because of the smaller pore size of the latter.

The relationship between the retention volume and the amount of liquid phase can be related to the partition and adsorption terms. The values of these terms and the adsorption constants can be calculated from the previously mentioned expression (1):

$$V_N/W_L = K_L' + (K_a \, \sigma_A + K_S \, \sigma_S) \, W_S/W_L \tag{1}$$

where:

$K_a \, \sigma_A + K_S \, \sigma_S$, the adsorption term, is the slope of the straight line in the zone of low proportion of the liquid phase.

The solute-support adsorption constants K_S, can be calculated from $(K_a \, \sigma_A + K_S \, \sigma_S)$, if we know the value of the adsorption term at the gas-liquid interface. As the values of $K_a \, \sigma_A$ are generally much lower than $(K_a \, \sigma_A + K_S \, \sigma_S)$ (see refs. 1,2)

Table IV Adsorption term $K_a \, \sigma_A + K_S \, \sigma_S$ (cm^3/g) and adsorption constant K_S (cm) of Chromosorb 101 and Chromosorb 102 coated with Ethofat; Specific surface area σ_S of Chromosorb 102 support

INJECTED COMPOUNDS	$K_a \, \sigma_A + K_S \, \sigma_S$ (cm^3/g)		$K_S \cdot 10^4$ (cm)	σ_S (m^2/g)
	0.4-2.6% Ethofat CHROMOSORB 101	0.3-4.2% Ethofat CHROMOSORB 102	CHROMOSORB 101 CHROMOSORB 102	CHROMOSORB 102
n-PENTANE	33	100	0.66	151
n-HEXANE	115	294	2.30	128
n-HEPTANE	297	–	5.90	–
n-OCTANE	–	–	–	–
METHANOL	11	14	0.22	64
ETHANOL	22	35	0.44	79
ISOPROPANOL	35	70	0.70	100
TERT-BUTANOL	46	90	0.92	98
n-PROPANOL	62	114	1.24	92
DIETHYL ETHER	38	102	0.76	134
DI-ISOPROPYL ETHER	158	396	3.16	125
METHYL FORMATE	15	29	0.30	97
ETHYL FORMATE	46	90	0.92	98
METHYL ACETATE	53	91	1.06	86
ETHYL ACETATE	124	254	2.48	102
ISOPROPYL ACETATE	225	–	4.50	–
METHYL PROPIONATE	143	240	2.86	84
ETHYL PROPIONATE	–	–	–	–
FORMALDEHYDE	12	14	0.24	58
ACETALDEHYDE	11	19	0.22	86
PROPIONALDEHYDE	42	72	0.84	86
ACETONE	49	66	1	66
METHYL ETHYL KETONE	101	208	2	104
METHYLAL	46	99	0.92	107
WATER	6	7	0.12	58

we can, at a first approximation, ignore the adsorption at the gas-liquid interface and calculate K_S once we know σ_S (the specific surface area of the support). The values of K_S shown in Table IV have been calculated assuming that σ_S for Chromosorb 101 is equal to 50 m^2/g(4). These obtained values are relatively high for non-polar compounds and low for polar compounds and this suggests a higher adsorption of the former compared with the latter, in agreement with the behaviour of Chromosorb 101 as an adsorbent.

The adsorption constants, K_S, for Chromosorb 102 can be identically calculated by using the value of σ_S indicated by the manufacturer (4) as equal to 300-400 m^2/g, but the K_S obtained in this way would be too much smaller than that obtained for Chromosorb 101. As only minor differences in their chemical nature exist between the surface of these adsorbents having identical monomer composition (both being styrene-divinylbenzene polymers) we think that it is better to assume that the adsorption constants should be alike. Thus, the differences in the adsorption terms between Chromosorb 101 and Chromosorb 102 are due to differences in the "active" surface areas. The available surface area of Chromosorb 102 can be obtained from the experimental values of the adsorption terms of Chromosorb 102 and the values of K_S for each compound previously calculated for Chromosorb 101. The surface areas obtained by this method are indicated in Table IV.

We can observe that the obtained areas of Chromosorb 102 available for the different compounds are lower than the surface area indicated by the manufacturer and they are not identical for several substances, showing greater values for the non-polar compounds (e.g. hydrocarbons and ethers) than for the others, while the small values correspond to water and other polar compounds.

The first observation could be explained considering that the manufacturer's values of the specific surface areas are calculated from adsorption measurements by the BET method using nitrogen that is a small molecule and can penetrate into the narrow pores of Chromosorb 102 support, but, in the case of the studied compounds, their larger size renders more difficult the entrance into the micropores. The differences obtained in the support surface areas can be also explained by the unlike compound size. In fact, we can point out that a different specific surface area exists for each studied compound. It is noted that for non-polar compounds this explanation is valid since while the molecular weight is small, the specific surface area of the support is larger (hydrocarbons and ethers) (Table IV), but it does not work in the case of polar compounds in a homologous series, which have minor specific surface areas, according to their polarity. That is to say, the smaller the size of the compound, the smaller is its σ_S value.

The unlike behaviour showed by the polar and non-polar compounds could be explained considering the interactions that exist between the support, the liquid phase and the injected compounds. Then, we think that the σ_S values would represent a function of the molar volume of the solute, the specific solubility parameters, and the percentage of the liquid phase. We can write a general expression:

$$\sigma_S = f\ (V_i,\ \delta,\ \tau) \tag{2}$$

Several authors (5, 6, 7, 8) have proposed general equations for describing the solution and adsorption systems, relating the energy of the process with the specific solubility parameters and the molar volumes. In the case of an adsorption system, the energy of adsorption ΔE^A (5) can be described as:

$$\Delta E^A = V_i\ (\delta^i_d \delta^a_d + \delta^i_o\ \delta^a_o + \delta^i_{in}\ \delta^a_d + \delta^a_{in}\ \delta^i_d + \sigma^i_a \delta^a_b + \delta^a_a \delta^i_b) \tag{3}$$

In the case of a solution system, they consider an energy of solution ΔE^S:

$$\Delta E^S = V_i\ \left[(\delta^j_T)^2 - 2\delta^i_d \delta^j_d - 2\delta^i_o \delta^j_o - 2\delta^j_{in} \delta^i_d - 2\delta^i_a \delta^j_b - 2\delta^j_a \delta^i_b - 2\delta^i_{in} \delta^j_d \right] \tag{4}$$

where: the superscripts are i=solute, j=liquid solvent and a=adsorbent.

We have used these general equations in order to obtain a new expression that could explain our chromatographic system. At very low percentages of the stationary phase when calculating the K_S value we must take into consideration two different interactions: the solute-adsorbent and the solute-stationary phase interaction. If we assume that at 23% of Ethofat on

Chromosorb 102, where there is a minimum in the graph of Fig. 2, all the support surface is covered, then at 4.2% of the liquid phase the contribution of Ethofat 60/25 on the retention will be only 0.18 (4.2/23). Thus, assuming that there exists a competition between the adsorption and the solution, and this last one makes more difficult the adsorption process, we propose a new equation that takes into account the percentage of contribution due to the solution. The proposed equation is:

$$\Delta E' = \Delta E^A - \gamma \Delta E^S \tag{5}$$

where $\Delta E'$ is the new adsorption energy of the process related to the surface area and the adsorption constant.

Using this equation, we have calculated $\Delta E'$ of 14 of the injected compounds of varying polarity, for which we have found the solubility parameters in the bibliography. We have to point out that these solubility parameters are not accurately determinated, and different authors list different values for the compound (6, 7, 8). The used values were:

1. For the solutes, the values proposed by R. Tijssen et al. (8).

2. For the solid support, Chromosorb 102, which is a styrene-divinyl benzene copolymer, we have considered that it has a similar structure to benzene and the same δ values. The most important term, in the δ values, is the dispersion solubility parameter, and it means that in equation 5 we have only considered the term $\delta_d^a \delta_d^i$.

3. For the liquid phase, Ethofat 60/25, we have not found the δ values in the bibliography, so we have used the δ values of a phase with a similar behaviour, a phthalate, with $\delta_T^j = 12.2$, $\delta_d^j = 11.8$, $\delta_b^j = 8.5$, $\delta_o^j = 1.71$. We did not take into account δ_{in}^j and δ_a^j because of their lower value compared with the other ones.

From equation 5 and the aforementioned considerations, we obtain a simplified expression:

$$\Delta E' = V_i \left[\delta_d^i \delta_d^a - \tau \left[(\delta_T^j)^2 - 2\delta_a^i \delta_b^j - 2\delta_d^i \delta_d^j - 2\delta_o^i \delta_o^j \right] \right] \tag{6}$$

Table V Values of $\sigma_S/\Delta E'$ for several compounds

INJECTED COMPOUNDS

n-PENTANE	0.018
n-HEXANE	0.014
METHANOL	0.019
ETHANOL	0.016
ISOPROPANOL	0.016
TERT-BUTANOL	0.014
n-PROPANOL	0.015
DIETHYL ETHER	0.019
METHYL ACETATE	0.016
ETHYL ACETATE	0.015
ACETONE	0.013
METHYL ETHYL KETONE	0.016
WATER	0.019
Average value	0.016

The values of $\sigma_S/\Delta E'$ are listed for these 14 compounds in Table V.

The relatively good agreement of these values seems to indicate that our considerations are correct and thus, the observed σ_S differences can be atrributed to the molar volumes and to the interactions between the solute and the phases, expressed as the solubility parameters. At present, we continue our investigations in this way and certainly the surface of Chromosorb 101 will be involved in similar considerations.

However, a part of the unlike values of the surface area obtained for each compound, the values indicated in Table IV permit us to obtain an average surface area for Chromosorb 102 that would be about 95 m^2/g, 30% of the surface area value indicated by the manufacturer which is in accordance with the value calculated by Gearhart and Burke (9) using the retention volumes obtained with the clean adsorbent.

This means that the "active" surface area of Chromosorb 102 is approximately double of the surface area of Chromosorb 101, which also coincides with the relationship between the percentage of the minimums of the graphs of $V_N/W_S = f(W_L/W_S)$, that is 10% for the Chromosorb 101 and 23% for the Chromosorb

102, what means that till these percentages of liquid phase the support action is important.

If the support acts even with such a relatively high amount of liquid phase, we must accept that it does not remain perfectly covered when the liquid phase is added. It seems as the support is not coated with a uniform monolayer, as pointed out by Giddings (10), but a portion of the support is not perfectly coated and this is responsible for the observed adsorption. We have noticed it by using scanning electron microscope (3) when we studied Chromosorb 101 coated with a liquid phase. As can be seen in Fig.3a and Fig.3b, where the outer and the inner surface of Chromosorb 101 coated with 20% of Ethofat are shown, the liquid phase is basically placed on the periphery of the support particle while its inner part remains relatively porous.

If we compare the surface of Chromosorb 101 and Chromosorb 102 (Figs 3c and 3d), we can observe that the former seems to be more porous than the latter. However, using a scanning electron microscope with higher resolution, magnification and accelerator voltage, we can observe that the surface of Chromosorb 102 having a homogeneous appearance (Fig. 3d) is now really inhomogeneous and very porous (Fig. 3e).

A similar comparison can be done concerning the appearance of the two supports coated with the liquid phase. For Chromosorb 102 with 18.8% Ethofat, Fig. 3f and Fig. 3g show the outer and the inner surface, respectively. The inner surface seems to be less porous as compared with the corresponding Chromosorb 101 (Fig. 3b). Also, if we use the JEOL microscope, we can observe a really porous outer surface as seen in Fig. 3h.

From these observations, we can point out that for both supports, even with large amounts of the liquid phase, the outer surface has always a certain degree of porosity showing small holes, and the inside of the particle shows an even more porous appearance than its outside.

In conclusion, we can say that the two adsorbents used in this work have the same behaviour. These results confirm the assumption expressed in our previous paper (3), i.e., that the stationary phase does not go mainly into the pores, as suggested

Fig. 3(a)

Fig. 3(b)

Fig. 3(c)

Fig. 3(d)

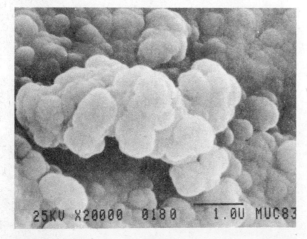

Fig. 3(e)

Fig. 3(f)

Fig. 3(g)

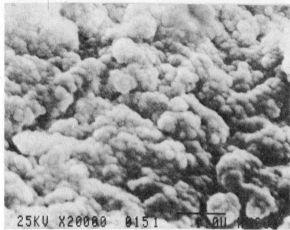

25KV X20000 0151   Fig. 3(h)

Fig. 3. Microphotographs of:

   (a) 20% Ethofat 60/25 on Chromosorb 101. Outer surface.
       (Magnification: 10.000 X).
   (b) 20% Ethofat 60/25 on Chromosorb 101. Inner surface.
       (Magnification: 10.000 X).
   (c) Chromosorb 101. (Magnification: 5000 X).
   (d) Chromosorb 102. (Magnification: 10.000 X).
   (e) Chromosorb 102. (Magnification: 20.000 X).
   (f) 18.8% Ethofat 60/25 on Chromosorb 102. Outer suface.
       (Magnification: 10.000 X).
   (g) 18.8% Ethofat 60/25 on Chromosorb 102. Inner suface.
       (Magnification: 10.000 X).
   (h) 18.8% Ethofat on Chromosorb 102. (Magnification:
       20.000 X).

by Kashtock (11), but it has a certain tendency to lie on the external surface of the particle. As can be seen in the photographs, the external surface shows a host of small holes through which the solute molecules can circulate and arrive to the inner surface of the support. This fact can explain the observed adsorption when the amount of liquid phase is as large as 10% on Chromosorb 101 and 23% on Chromosorb 102.

ACKNOWLEDGEMENTS

The authors gratefully acknowledge the skilful technical assistance of the Servei de Microscopia of the Universitat de Barcelona, especially of Mr. Fontarnau, and of the Department of Metallurgy of the Universidad Complutense de Madrid.

LIST OF SYMBOLS

V_N : net retention volume

K_a : adsorption constant at gas liquid interface

K_S : adsorption constant of support

K'_L : experimental partition constant of liquid phase

A_L : surface area of liquid phase in the column

A_S : surface area of support in the column

$\sigma_A = A_L/W_S$ specific surface area of the liquid phase

$\sigma_S = A_S/W_S$ specific surface area of the support

W_L : weight of the liquid phase in the column

W_S : weight of the support in the column

V_i : molar volume of the injected compound

δ : specific solubility parameter

τ : percentage of effective stationary phas

δ_T : total solubility parameter

714

δ_d : dispersion solubility parameter

δ_o : orientation solubility parameter

δ_{in}: induction solubility parameter

δ_a : proton donor solubility parameter

δ_b : proton acceptor solubility parameter

ΔE^A: energy of adsorption

ΔE^S: energy of solution

$\Delta E'$: corrected adsorption energy

REFERENCES

1. D. Barceló, M.T. Galcerán and L. Eek, Chromatographia 12 725 (1979

2. D. Barceló, M.T. Galcerán and L. Eek, Chromatographia 14 73 (1981

3. D. Barceló, M.T. Galcerán and L. Eek, J. Chromatogr. 217 109 (1981)

4. Johns Manville. Chromosorb century series. FF-202-A.

5. R.A. Keller, B.L. Karger and L.R. Snyder, in R. Stock (Ed.), Gas Chromatography 1970, Institute of Petroleum, London, 1971, p. 125.

6. B.L. Karger, L.R. Snyder and C. Eon, Anal. Chem. 50 2126 (1978)

7. B.L. Karger, L.R. Snyder and C. Eon, J. Chromatogr. 125 71 (1976)

8. R. Tijssen, H.A.H. Billiet, and P.J. Schoenmakers, J. Chromatogr., 122 185 (1976)

9. H.L. Gearhart and M.F. Burke, J. Chromatogr. Sci. 11 411 (1973)

10. J.C. Giddings, Anal. Chem. 34 458 (1962)

11. M.E. Kashtock, J. Chromatogr. 176 25 (1979)

CALCULATION AND OPTIMIZATION METHODS

CHROMATOGRAPHIC PARAMETERS OF LIPOPHILICITY OF A SERIES OF DERMORPHIN RELATED OLIGOPEPTIDES

G.L. BIAGI, A.M. BARBARO, M.C. GUERRA, G. CANTELLI FORTI, P.A. BOREA*, M.C. PIETROGRANDE**, S. SALVADORI*** and R. TOMATIS***

Istituto di Farmacologia, Università di Bologna, Bologna, Italy
 *Istituto di Farmacologia, Università di Ferrara, Ferrara, Italy
 **Istituto di Chimica, Università di Ferrara, Ferrara, Italy
***Istituto di Chimica Farmaceutica, Università di Ferrara, Ferrara, Italy

Ever since the work of Overton (1,2), Meyer (3,4), Baum (5), Ferguson (6), Collander (7,8), Zahradnik (9) and Hemker (10) it has been recognized that the lipophilic character of molecules plays an important role in determining their biological activity. The work of Hansch and coworkers obtained numerically defined constants to assess hydrophobic character (11, 12). Today log P is generally used as an expression of the lipophilic character of molecules, the shake—flask procedure being considered the standard reference procedure for the determination of P, which is the octanol/water partition coefficient. However this technique suffers from several problems. It is time consuming and subject to purity, stability and mass balance problems. According to a first suggestion by Boyce and Milborrow (13), as a shortcut to hydrophobic constants in our laboratory the R_m values have been measured by means of reversed-phase thin—layer chromatography (14, 15); more recently we have also investigated the suitability in QSAR of log k' values obtained by means of reversed—phase HPLC (16). The discovery of enkephalins and endorphins with high affinities for opioid receptors had added new dimensions to the study of the mechanism of action of opioids. In recent times there has been a growing interest in the study of structure—activity relathionships of opioid agonists (17, 18, 19, 20). The present study was undertaken in order to determine the R_m and log k' values of a series of 23 dermorphin derivatives which had been synthetized previously (21, 22). Their correlation with octanol/water partition data in view of QSAR studies was pointed out.

MATERIALS AND METHODS
Determination of R_m values by means of reversed—phase TLC

The details of the chromatographic technique have been described previously (15). Glass plates measuring 20 x 20 cm were coated with Silica Gel G in the usual manner (14). However in order to obtain a better control of the pH of the stationary phase the slurry of Silica Gel G was obtained with 0.09 N NaOH. A stationary non—polar phase was obtained by impregnating the Silica Gel G layer with Silicone DC 200 (350 cS) from Applied Sciences Laboratories. The impregnation was carried out by developing the plates in a 5% silicone solution in ether. Eight plates could be impregnated, in a single chromatographic chamber, containing 200 ml of the silicone solution. The plates were left in the chamber for 12 h, that is for several hours after the silicone solution had reached the top of the plates. This method of impregnating Silica Gel G layers had also been used by Roomi et al. (23) with silicone oil, by Boyce and Milborrow (13) with liquid paraffin and by several other investigators as reported by Kirchner (24), Anker et al. (25) and Seydel et al. (26). In order to avoid the "edge effect" and uneven migration of the solvent front, the silicone impregnated layer was cut 2 cm from each lateral edge and the chromatographic chamber was saturated with the mobile phase vapor as indicated by Stahl (14).

719

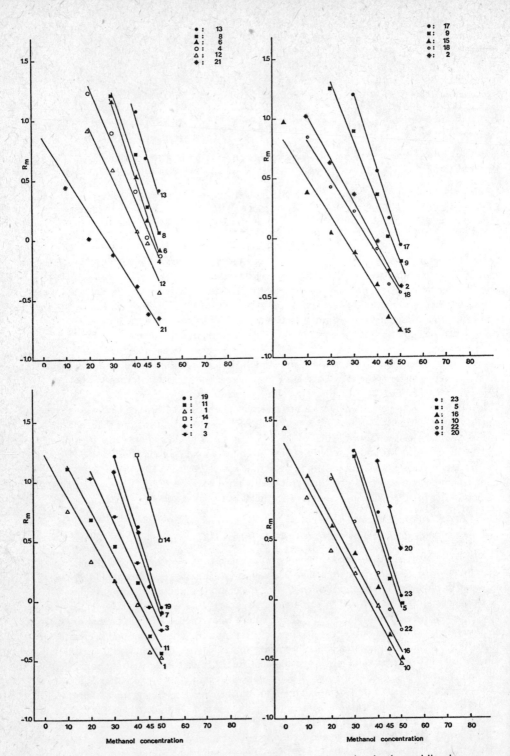

Fig. 1. Relationship between R_m values and methanol concentration in the mobile phase.

A migration of 10 cm was obtained on all the plates by cutting the layer at 12 cm and spotting the compounds on a line 2 cm from the lower edge of the plate. The mobile phase saturated with silicone was an aqueous buffer (sodium acetate — Veronal buffer 1/7 M at pH 7.0) alone or mixed with various quantities of methanol. Two plates were developed simultaneously in a chromatographic chamber containing 200 ml of mobile phase. The dermorphin derivatives were dissolved in methanol (1–2 mg/ml) and 1 μl of solution was spotted on the plates in randomized allocations in order to avoid any systematic error. The developed plates were dried and sprayed with an alkaline solution of potassium permanganate. After a few minutes at 120°C, yellow spots appeared on an intensely pink background. The R_m values were calculated by means of the formula:

$$R_m = \log \left(\frac{1}{R_f} - 1 \right)$$

where $R_f = \dfrac{\text{distance of sample spot from starting point}}{\text{distance of solvent front from starting point}}$

Determination of log k' values by means of reversed—phase HPLC

Chromatography was performed on a Waters 6000 A chromatograph using a μBondapack C_{18} column (300 x 3.9 mm I.D.) (Waters), packed with Silica Gel (particle size 10 μm) with a C_{18} chemically bonded non—polar stationary phase. A UV detector (Waters Model 480) at 214 nm and Hamilton 802 chromatographic syringes (25 μl) were also used. The dermorphin derivatives were separated using methanol—water mixtures as the mobile phase at a flow—rate of 1 ml/min. The methanol concentration ranged from 60 to 80% . Samples were dissolved in methanol (1 mg/ml) and applied to the column in 5 μl volumes. All solutions were first filtered to reduce contamination. The experiments were performed at room temperature (20–22°C). The retention times were expressed as log capacity factor (k') where $k' = (t_x - t_o) / t_o$.

RESULTS
R_m values

The spraying of the developed plates with potassium permanganate resulted in the appearance of round spots at different distances from the starting line. At 0% methanol in the mobile phase for only a few compouneds, i.e. the most hydrophilic ones, it was allowed to determine a reliable R_f value. The most lipophilic compounds remained close to the origin. Therefore in order to obtain suitable R_f values it was necessary to add methanol to the mobile phase. The transformation of the R_f values into R_m values yielded the data reported in Table 1. Higher and/or positive R_m values indicate compounds more lipophilic than those represented by a lower and/or negative R_m value. The plots of Fig. 1 show that for each dermorphin derivative there is a linear relationship between the R_m values and the composition of the mobile phase over a particular range of methanol concentration. The straight lines in Fig. 1 were calculated by means of the least squares method from the R_m values in the linearity range (see Table 1). The theoretical R_m values at 0% methanol in the mobile phase, i.e. in a standard system where all the compounds could be compared, are represented by the intercept of the equations reported in Table 2.

Equation (1) shows that there is a highly significant linear correlation between the in-

TABLE 1. R_m values of dermorphin - related oligopeptides at increasing methanol concentrations

Cpd. No.	Methanol Concentration (%)										Range of linearity
	0	10	20	30	40	45	50	60	70	80	
1	1.40	0.76	0.33	0.17	−0.03	−0.43	−0.48	−0.51	−0.66	−	0−50
2	−	1.02	0.63	0.36	−0.03	−0.28	−0.41	−0.53	−0.72	−	10−50
3	−	−	1.03	0.70	0.32	−0.05	−0.25	−0.38	−0.56	−0.50	20−50
4	−	−	1.23	0.89	0.40	0.02	−0.14	−0.45	−0.76	−	20−50
5	−	−	−	1.20	0.57	0.17	−0.03	−0.36	−0.74	−	30−50
6	−	−	−	1.16	0.51	0.15	−0.10	−0.36	−0.69	−	30−50
7	−	−	−	1.08	0.58	0.12	−0.10	−0.29	−0.64	−	30−50
8	−	−	−	1.20	0.71	0.27	0.05	−0.24	−0.61	−	30−50
9	−	−	1.25	0.89	0.36	0.02	−0.20	−0.43	−0.68	−	20−50
10	1.44	0.85	0.40	0.21	−0.06	−0.42	−0.54	−0.57	−0.71	−	0−50
11	−	1.12	0.68	0.46	0.15	−0.30	−0.45	−0.48	−0.65	−	10−50
12	−	−	0.92	0.58	0.09	−0.29	−0.45	−0.64	−	−	20−50
13	−	−	−	−	1.07	0.68	0.40	−0.06	−0.46	−0.55	40−50
14	−	−	−	−	1.22	0.86	0.52	−0.04	−0.34	−0.37	40−50
15	0.98	0.39	0.05	−0.13	−0.39	−0.66	−0.78	−0.76	−	−	0−50
16	−	1.03	0.61	0.38	0.10	−0.30	−0.50	−0.58	−	−	10−50
17	−	−	−	1.20	0.57	0.17	−0.06	−0.35	−0.67	−	30−50
18	1.52	0.84	0.43	0.23	0.09	−0.39	−0.46	−0.56	−0.63	−	10−50
19	−	−	−	1.22	0.62	0.26	−0.05	−0.33	−0.75	−	30−50
20	−	−	−	−	1.15	0.77	0.42	−0.01	−0.46	−0.48	40−50
21	1.06	0.44	0.01	−0.12	−0.39	−0.63	−0.66	−0.74	−	−	0−50
22	−	−	1.01	0.65	0.22	−0.09	−0.26	−0.49	−	−	20−50
23	−	−	−	1.24	0.47	0.34	0.03	−0.28	−0.68	−	30−50

tercepts and the slopes of the straight lines (Table 2) describing the relationship between R_m values and methanol concentration in the mobile phase.

$$R_m = -1.313 + 72.12 \, b \qquad \begin{matrix} n \\ 23 \end{matrix} \quad \begin{matrix} r \\ 0.990 \end{matrix} \quad \begin{matrix} s \\ 0.148 \end{matrix} \, (1)$$
$$(F = 1033.40; P < 0.005)$$

According to Draffehn et al. (27) who investigated a series of steroids, the straight lines relating R_m values and methanol concentration in the mobile phase fall more steeply the more hydrophobic is the molecule. In particular the slope of the straight line is due to fact that the decrease in the proportion of water in the mobile phase provokes a decrease in the number of water contacts in the mixed solvent envelope around the molecule. On the other hand the decrease in the number of water contacts is dependent on the lipophilicity of the molecule. Draffehn et al. (27) have also proposed the area under the straight line describing the relationship between R_m values and organic solvent concentration as a measure of hydrophobicity that is independent of the portion of the organic component in the mobile phase. In Table 2 are reported the areas under the straight lines described by the equations of Table 2 within the limits from 0 to 50% methanol in the mobile phase.

TABLE 2 . R_m , $\Sigma\pi$ and F_D values of dermorphin - related oligopeptides.

H - CH - CH - C - NH - CH - C - NH - CH - C - NH - R_1 - C - R_2

Tyr | D - Ala | Phe | Gly or β - Ala

Cpd. No.	R_1	R_2	TLC equation R_m =a	b	r	$\Sigma\pi$	F_D
1	CH_2	$N\begin{smallmatrix}H\\H\end{smallmatrix}$	1.22	−0.035	0.979	−1.23	0.619
2	CH_2	$N\begin{smallmatrix}H\\CH_2CH_3\end{smallmatrix}$	1.38	−0.036	0.998	0.08	0.694
3	CH_2	$N\begin{smallmatrix}CH_2CH_3\\CH_2CH_3\end{smallmatrix}$	1.95	−0.043	0.991	1.18	0.980
4	CH_2	$N\begin{smallmatrix}H\\CH_2\text{-}\bigcirc\end{smallmatrix}$	2.24	−0.048	0.993	0.78	1.126
5	CH_2	$N\begin{smallmatrix}H\\CH_2CH_3\text{-}\bigcirc\end{smallmatrix}$	3.09	−0.063	0.996	1.34	1.553
6	CH_2	$N\begin{smallmatrix}H\\CH\\CH_3\end{smallmatrix}-\bigcirc$	3.07	−0.064	0.999	1.34	1.543
7	CH_2	$N\begin{smallmatrix}CH_3\\CH_2\text{ -}\bigcirc\end{smallmatrix}$	2.93	−0.061	0.992	1.34	1.273
8	CH_2	$N\begin{smallmatrix}CH_3\\CH_2CH_2\text{-}\bigcirc\end{smallmatrix}$	3.00	−0.059	0.993	1.90	1.507
9	CH_2	$N\begin{smallmatrix}H\\\end{smallmatrix}\diagdown$	2.30	−0.050	0.995	0.91	1.156
10	CH_2	$N\begin{smallmatrix}H\\CH_2CH_2OH\end{smallmatrix}$	1.31	−0.037	0.988	−0.59	0.662
11	CH_2	$N\begin{smallmatrix}H\\CH_2CH_2OCH_3\end{smallmatrix}$	1.52	−0.038	0.985	0.06	0.765
12	CH_2	$N\begin{smallmatrix}H\\CH_2CH_2\text{-}\bigcirc\text{-OH}\end{smallmatrix}$	1.85	−0.044	0.989	0.67	0.930

TABLE 2.

Cpd. No.	R_1	R_2	TLC equation $R_m = a$	b	r	$\Sigma \pi$	F_D
13	CH_2	N–H adamantyl	3.73	−0.067	0.996	2.14	1.873
14	CH_2	N–H CH_2 - adamantyl	4.02	−0.070	1.000	2.70	2.019
15	CH_2	OH	0.83	−0.032	0.984	−0.67	0.419
16	CH_2	O - CH_2CH_3	1.41	−0.037	0.987	0.38	0.710
17	CH_2	O - CH_2 - ⬡	3.13	−0.065	0.997	1.66	1.573
18	$(CH_2)_2$	N–H, H	1.15	−0.032	0.993	−0.95	0.579
19	$(CH_2)_2$	N–H, $CH–$⬡, CH_3	3.15	−0.064	0.999	1.62	1.583
20	$(CH_2)_2$	N–H adamantyl	4.06	−0.073	1.000	2.42	2.039
21	$(CH_2)_2$	OH	0.86	−0.032	0.977	−0.39	0.434
22	$(CH_2)_2$	O - CH_2CH_3	1.91	−0.043	0.997	0.66	0.960
23	$(CH_2)_2$	O - CH_2 - ⬡	3.09	−0.061	0.997	1.94	1.553

The areas under the straight lines were calculated by means of the equation

$$F_D = \frac{4a - b}{8} \qquad (2)$$

obtained by solving the integral between 0 and 50% methanol in the mobile phase and where a and b are the coefficients of the equations reported in Table 2.

However because of eq. 1 and 2 one should have expected the very high correlation coefficient shown by eq. 3 between F_D and R_m values.

$$F_D = 0.007 + 0.496\ R_m \qquad\qquad \begin{array}{ccc} n & r & s \\ 23 & 0.997 & 0.042\ (3) \end{array}$$
$$(F = 3134.35; P < 0.005)$$

Therefore at least in the present series of compounds the F_D values should have the same meaning of the R_m values. The data reported in Table 2 show the strong lipophilic character of the adamantyl group. In fact cpds. no. 13, 14 and 20 with such a substituent are the most lipophilic ones. In particular cpds. 13 and 20 are 323 and 813 times respectiverly more hydrophobic than cpds. no. 1 and 18. In Table 3 are reported the ΔR_m values showing the lipophilicity of β-alanine vs. glycine and adamantyl group compared with π values.

TABLE 3. ΔR_m, $\Delta \log k'$ and π values of β-alanine vs. glycine and adamantyl group.

Group	Cpd./Cpd.	ΔR_M	$\Delta \log k'$	π [a]
β-alanine vs. glycine	18 - 1; 19-6; 20 - 13;21-15; 22 -16;23-17;	0.14	0.09	0.25
Adamantyl	13 - 1; 20 - 18;	2.71	2.78	3.39

a : See ref. 28 .

Relationship between R_m and $\Sigma\pi$ values

In Table 2 are reported the $\Sigma\pi$ values calculated from the data of Hansch et al. (12) and Fauchère et al. (28). In particular the π value of the adamantyl group as reported in Table 3 was obtained in the following manner:

π adamantyl = log P adamantylalanine − log P alanine = 0.43 −(−2.96) = 3.39

As regards the contribution of the CH_2 group(s) at the R position the $\Sigma\pi$ values were calculated by considering only one CH_2 group in the β−alanine compounds and none in the glycine derivatives.

Equation 4 shows a very high correlation coefficient between R_m and $\Sigma\pi$ values.

$$R_m = 1.586 + 0.867\ \Sigma\pi$$
$$(F = 174.48; P < 0.005)$$

	n	r	s
	23	0.945	0.333 (4)

Obvioulsy because of eq. 3, eq. 5 shows a highly significant relationship between $\Sigma\pi$ and F_D values.

$$F_D = 0.792 + 0.430\ \Sigma\pi$$
$$(F = 171.25; P < 0.005)$$

	n	r	s
	23	0.944	0.170 (5)

Log k' values and their relationship with R_m and $\Sigma\pi$ values

The HPLC procedure allowed to show that there is a linear relationship between log k'

725

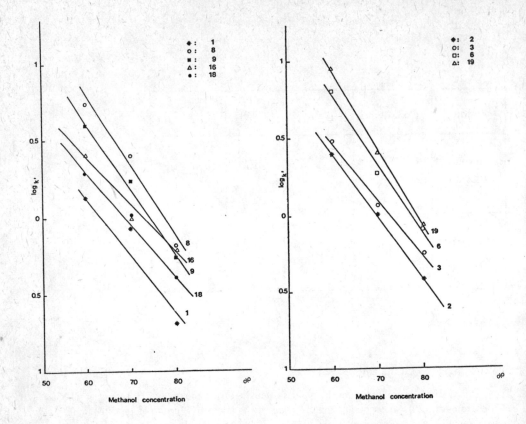

Fig. 2. Relationship between log k' values and methanol concentration in the mobile phase.

values and methanol concentration in the mobile phase. In Fig. 2 some of the data were reported. The equations of all the straight lines were reported in Table 4. However it must be pointed aut that for cpds. no. 13, 14, 15, 17, 20 and 23 only two experimental points could be obtained. In fact while cpd. no. 15 was eluted with the solvent front at 80% methanol in the mobile phase, cpds. no. 13, 14, 17, 20 and 23 were not eluted at 60% methanol in the mobile phase. Therefore the equations for the above compounds were calculated with only two data points and obviously had a correlation coefficient of 1.000. However despite this lack of accuracy in calculating six of the equations of Table 4, eq. 6 yields a very high correlation coefficient between the intercepts and the slopes of Table 4. On the other hand eq. 7 calculated without cpds. no. 13, 14, 15, 17, 20 and 23 is very similar to eq. 6.

	n	r	s
$\log k' = -0.662 + 90.03\ b$	23	0.986	0.196 (6)
$(F = 764.48; P < 0.005)$			

726

TABLE 4. HPLC data of dermorphin derivatives

Cpd. No.	HPLC equation		
	log k′ =a	b	r
1	2,35	− 0,037	0.927
2	2,86	− 0.041	0.999
3	2,69	− 0.037	0.999
4	3,56	− 0,049	0.998
5	3,87	− 0,052	0.998
6	3,48	− 0.045	0.996
7	4,03	− 0,052	0.999
8	3,57	− 0.046	0.990
9	3,20	− 0,043	0.996
10	1,40	− 0,024	0.999
11	1,77	− 0,027	0.997
12	3,43	− 0,050	0.999
13	5,24	− 0,065	1.000
14	4.61	− 0,055	1.000
15	1,18	− 0,022	1.000
16	2,26	− 0,031	0,985
17	4,30	− 0,056	1.000
18	2,35	− 0,034	0,996
19	3.99	− 0,051	0,999
20	4,82	− 0,057	1.000
21	0,88	− 0,013	0,989
22	3,30	− 0,044	0,998
23	4.04	− 0,051	1.000

$$\log k' = -0.403 + 82.60 \, b \qquad n = 17 \qquad r = 0.985 \qquad s = 0.166 \ (7)$$
$$(F = 486.14; P < 0.005)$$

Equations 6 and 7 show that the linear relationship between chromatographic data and methanol concentration in the mobile phase holds both in the reversed–phase TLC and HPLC systems.

The relationship between log k′ values and both R_m and $\Sigma\pi$ values is shown by eq. 8 and 9.

$$R_m = -0.199 + 0.786 \log k' \qquad n = 23 \qquad r = 0.928 \qquad s = 0.379 \ (8)$$
$$(F = 130.34; P < 0.005)$$

$$\log k' = -1.771 + 0.820 \, \Sigma\pi \qquad n = 23 \qquad r = 0.871 \qquad s = 0.557 \ (9)$$
$$(F = 65.88; P < 0.005)$$

Equations 8 and 9 show correlation coefficients somewhat lower and standard deviations somewhat higher than those yielded by eq. 4. However the overall result is acceptable when considering that the log k' values where extrapolated from higher methanol concentrations by means of equations calculated with only 2 or 3 data points. The Δ log k' values for β—alanine vs. glycine and the adamantyl group as reported in Table 4 are quite similar to the Δ R_m values.

DISCUSSION AND CONCLUSION

In the present series of dermorphin derivatives the R_m values from TLC and the log k' values from HPLC seem to be very well correlated between them and with the π values calculated according to Hansch et al. (12).

All this could be very useful in QSAR studies. On the other hand the present data seem to allow some more general considerations. The linear correlation between the intercepts and the slopes of the TLC or HPLC straight lines could be used in order to check the cromatographic data. In fact deviations from the linearity in eq. 1 and 7 should indicate that the TLC or HPLC equations had been calculated out of the range of linearity between the R_m or log k' values and the methanol concentration in the mobile phase.

The very similar slopes of eq. 4 and 9 show that the change in lipophilicity indicated by the π values yields the same variation in R_m and log k' values. In other words the TLC and HPLC systems are equally sensitive to the variations of lipophilicity of molecules as expressed by π values. A different choice of the experimental conditions could lead to closer intercepts of the TLC and HPLC straight lines.

In conclusion, while the classical method for determining the octanol/water partition coefficient is likely to give more accurate and unequivocal data, the chromatographic technique seems to have several advantages: (a) it is simple and rapid; (b) it requires little material; (c) the material does not need to be very pure because impurities are separated during the determination; (d) the detection of spots by unspecific methods avoids the need for specific quantitative analytical methods; (e) the determination of the partition coefficient of slightly water—soluble compounds requires a long period of equilibration to achieve through partitioning between the phases.

However more work should be necessary in order to find standard conditions closer to those provided by the classical shake—flask method.

REFERENCES

1. E. Overton, Ueber die osmotischen Eigenschaften der Zelle in ihrer Bedeutung für die Toxicologie und Pharmakologie. Z. Physikal, Chem., 22, 189, 1897.

2. E. Overton, Studien über die Narkose. G. Fischer, Jena, 1901.

3. H. Meyer, Zur Theorie der Alkoholnarkose. Erste Mitteilung. Welche Eigenschaft der Anästhetica bedingt ihre narkotische Wirkung? Arch. Exp. Pathol. Pharmakol., 42, 109, 1899.

4. H. Meyer, Zur Theorie der Alkoholnarkose. Dritte Mitteilung. Der Einfluss wechselnder Temperatur auf Wirkungsstärke und Teilungscoefficient der Narkotica. Arch. Exp. Pathol. Pharmakol., 46, 338, 1901.

5. F. Baum, Zur Theorie der Alkoholnarkose. Zweite Mitteilung. Ein physikalisch-chemischer Beitrag zur Theorie der Narkotica. Arch. Exp. Pathol. Pharmakol., 42, 119, 1899.

6. J. Ferguson, The use of chemical potentials as indices of toxicity. Proc. Roy. Soc., Ser. B127, 787, 1939.

7. R. Collander, The partition of organic compounds between higher alcohols and water. Acta Chem. Scand., 5, 774, 1951.

8. R. Collander, The permeability of Nitella cells to non—electrolytes. Physiol. Plant., 7, 420, 1954.

9. R. Zahradnik, Influence of the structure of aliphatic substituents on the magnitude of the biological effect of substances. Arch. Int. Pharmacodyn. Thér., 135, 311, 1962.

10. H.C. Hemker, Lipid solubility as a factor influencing the activity of uncoupling phenols. Biochim. Biophys. Acta, 63, 46, 1962.

11. A.J. Leo, C. Hansch and D. Elkins, Partition coefficients and their uses. Chem. Rev., 71, 525, 1971.

12. C. Hansch and A.J. Leo, Substituent constants for correlation analysis in chemistry and biology. John Wiley & Sons, New York, 1979.

13. C.B.C. Boyce and B.V. Milborrow, A simple assessment of partition data for correlating structure and biological activity using thin-layer chromatography. Nature, 208, 537, 1965.

14. E.G. Stahl, Thin—layer chromatography. Springer, Berlin, 1969. •

15. G.L. Biagi, A.M. Barbaro, M.F. Gamba and M.C. Guerra, Partition data of penicillins determined by means of reversed—phase thin—layer chromatography. J. Chromatogr., 41, 371, 1969.

16. M.C. Guerra, A.M. Barbaro, G. Cantelli Forti and G.L. Biagi, R_m values, retention times and octanol—water partition coefficient of a series of 5—nitroimidazoles. J. Chromatogr., 259, 329, 1983.

17. J.S. Morley, Structure—activity relationships of enkephalin—like peptides. Ann. Rev. Pharmacol. Toxicol., 20, 81, 1980.

18. F.A. Gorin, T.M. Balasubramanian, T.J. Cicero, J. Schwietzer and G.R. Marshall, Novel analogues of enkephalin: identification of functional groups required for biological activity. J. Med. Chem., 23, 1113, 1980.

19. Kim Quang Do, J.L. Fauchère and R. Schwyzer, Electronic, steric and hydrophobic factors influencing the action of enkephalin—like peptides on opiate receptors. Hoppe—Seyler's Z. Physiol. Chem., 362, 601, 1981.

20. L. Nodasdi, D. Yamashiro, Chols Hao Li and P. Huidobro—Taro, Enkephalin analogs: synthesis and properties of analogs with lipophilic or extended carboxyl—terminus. Quantitative structure—activity relationship of analogs modified in residue position 5. Int. J. Peptide Res., 21, 344, 1983.

21. S. Salvadori, G.P. Sarto and R. Tomatis, Synthesis and pharmacological activity of dermorphin—tetrapeptide analogs. Eur. J. Med. Chem., in press, 1983.

22. G.P. Sarto, P.A. Borea, S. Salvadori and R. Tomatis, Pharmacological studies of a series of dermorphin related tetrapeptides. Arzneim. Forsch., in press, 1983.

23. M.W. Roomi, M.R. Subboran, K.T. Achaya, Separation of fatty acetylenic, ethylenic and saturated compounds by thin—layer chromatography. J. Chromatogr., 16, 106, 1964.

24. J.G. Kirchner, Thin—layer chromatography. Wiley—Interscience, New York, 1978.

25. L. Anker, D. Sonanini, Beitrag zur dünnschichtchromatographischen Prüfung von Fetten und Fetten Olen der Pharmakopöe Pharm. Acta Helv., 37, 360, 1962.

26. J.K. Seydel, K.J. Schaper, Chemische Struktur und biologische Aktivität von Wirkstoffen. Verlag Chemie, New York, 1979.

27. J. Draffehn, K. Ponsold, B. Schönecker, R_m values of steroids as an expression of their hydrophobic character. J. Chromatogr., 216, 69, 1981.

28. J.L. Fauchère, Kim Quang Do, P.J.C. Jow and C. Hansch, Unusually strong lipophilicity of "fat" or "super" amino acids, including a new reference value for glycine. Experientia, 36, 1203, 1980.

COMPARISON OF OPTIMIZATION METHODS FOR A SEPARATION QUALITY PARAMETER IN HPLC

JOSEP RAFEL

S.A. Cros, Research Center, Badalona (Barcelona), Spain

SUMMARY

In this work, the relation $Q = \dfrac{R}{t^{0.25}}$ (1) as a separation quality parameter (where R = Resolution calculated through the usual Purnell (1) equation, and t = retention time of the second peak.) has been used and optimized to improve the separation of peak pairs in a relatively short analysis time, not previously arbitrarily chosen, in samples containing nitro-aromatic compounds.

The optimization has been carried out through the use of three different methods which were finally compared, following one starting factorial design.

INTRODUCTION

Jandera and Churáček recently reviewed the theoretical retention characteristics in chromatography (2) using binary mobile phases, including detailed equations for resolution and optimization of the analytical conditions.

Wegscheider et al. obtained a chromatographic response function named CRT for automated optimization of separation (3).

This program calculates the expected retention time for the second one in the considered pair of peaks, using the optimal resolution value calculated in a previous work (5).

On the other hand in previous works (4, 5), "off-line" empirical resolution optimization was carried out after using a factorial design (6-8) followed by several different optimization methods such as a Simplex (9), a Box-Wilson (10) and a generalized Hooke-Jeeves method (11) in a FORTRAN program. In the present paper, a separation quality parameter, Q, is proposed for the the empirical evaluation of the achieved resolution between peak pairs; the results obtained are compared with values obtained using other methods. The optimal Q values give optimal resolutions in a short analysis time, not previously fixed.

EXPERIMENTAL
Material and Methods

All chemicals and solvents were the purest grades available from standard commercial sources. The nitroaromatic standards were obtained from Merck (Darmstadt, FRG).

Solutions containing nitrobenzene, 3,4-dinitrotoluene, 2-nitrotoluene and 3-nitrotoluene at individual concentrations around 0.01% were used.

Liquid Chromatography

A Hewlett-Packard Model 1080 Liquid Chromatograph, equipped with a variable-wavelength U.V. detector was used in our work. The column employed was obtained from Supelco: it was packed with LC-8 reversed-phase packing (5 μm).

Water-methanol mixtures were used as the eluent.

Results

First, a two-level factorial design was performed to study the influence of the eluent composition, system temperature, and flow rate on the separation quality parameter Q. The selected lower and upper coded limits are presented in Table I and the experiment matrix in Table II.

732

Table I Chosen lower (-1) and upper (+1) limits in the factorial design

i		x'_1	-1	+1
1	Methanol % (B)		40	60
2	Temperature (T)		35	70
3	Flow rate (F)		1	2

Table II Experiment matrix. Initial 2^3 factorial design

x'_1	x'_2	x'_3	Q
-1	-1	-1	1.41
+1	-1	-1	1.34
-1	+1	-1	1.41
+1	+1	-1	0.55
-1	-1	+1	1.64
+1	-1	+1	1.13
-1	+1	+1	1.12
+1	+1	+1	0.63

Result: $Q = 1.15 - 0.24 x'_1 - 0.23 x'_2 - 0.02 x'_3$ {2}

In the next step the three earlier mentioned optimization methods were applied and compared:

(a) The starting points of the Simplex optimization and the results obtained are presented in Table III.

(b) Data obtained in the Box-Wilson method are presented in Table IV.

(c) As indicated in Table V, in the FORTRAN program that extends the application of the Hooke-Jeeves method is run with new selected limit values for the variables. The optimal conditions and an expected value for the optimal Q obtained are given in Table VI.

Table III   Simplex method

Starting conditions: Methanol % (B) =40; Increase D = ± 5
       Temperature °C (T) =35   Increase D = ± 5
      (Flow rate is fixed at 1.5 ml/min)

Experiment matrix:

x'_1	x'_2	Q	Simplex
O	O	1.69	a b c d
0.9659	0.2591	1.57	a b - -
0.2591	0.9659	1.44	a - - -
0.7068	-0.7068	1.54	- b c d
0.2591	-0.9659	1.54	- - c -
O	O	1.71	- - - - e
-0.9659	-0.2591	1.56	- - - d

Max. age reached// optimal Q value Q = 1.70 achieved for:
      B = 40% and
      T = 35 °C when F maintained at 1.5 ml/min.

----

e = Duplication experiment done when min. age is reached/
  Expected Q value obtained.

   Van der Wal and J. Hubber (12) have calculated "Expected
retention time" values related to certain resolution values.

DISCUSSION

   Since resolution of 1.25 is usually high enough to have a
good qualitative and quantitative analysis and preparative
separations (13), the point is not the maximization of resolution itself, but the optimization of a practical chosen function which includes resolution and retention time. As a result,
a new enhanced method to obtain optimal conditions for HPLC

Table IV Experiment matrix and optimal conditions in the
Box–Wilson method

x'_1	u	x'_2	x'_3	Q
-0.4	0.6	-0.38	-0.03	1.14
-0.6	0.4	-0.58	-0.05	1.34
-0.8	0.3	-0.77	-0.07	1.34
-1	0.24	-0.96	-0.08	1.64
-1.2	0.2	-1.15	-0.10	1.74
-1.4	0.17	-1.35	-0.12	1.62
-1.6	0.15	-1.53	-0.13	1.46

Optimal Q obtained value

Q = 1.74

For: B = 38% Methanol

T = 32 $^{\circ}$C

F = 1.45 ml/min

Table V New chosen limit values for the extended Hooke–Jeeves
optimization

	Lower	Upper
x'_1	-1.25	1.5
x'_2	-1.125	1.5
x'_3	-1	1.5

Table VI   Optimal conditions in the Hooke-Jeeves optimization

% Methanol	37.5%
Temperature	33°C
Flow rate	1 ml/min

Expected Q: 1.72
Actual   Q: 1.67

Table VII   Retention-time forecast

Expected ret. time: 12.10
Actual value obtained: 12.87

analysis is suggested through the use of a quality factor
$Q = R/t^{0.25}$.

CONCLUSIONS

Although the optimal Q values obtained for the three
tested optimization methods are within the range of $\bar{Q} \pm 2\sigma$;
when the optimal conditions are compared in each method with
the mean and standard deviation of the others, the Hooke-Jeeves
method appears to be, as demonstrated in a previous work (5),
an easy and better way to achieve the optimal conditions
showing a greater agreement as indicated in Table VIII.
Due to the Q dependence on R and t, an estimation for the
retention time of the second peak using the optimal conditions
is obtained.

Table VIII Results of comparison

	S	H	M	B	D_1	W S	W H	W B
Q	1.70	1.67	1.70	1.74	0.04	+	+	+
B	40	37.5	38.5	38	1.32	+	+	+
T	35	33	33.3	32	1.53	+	+	+

	N	D_2	U	P	D_3	Y	T	D_4	Z
Q	1.71	0.01	+	1.72	0.03	+	1.69	0.02	-
B	37.75	0.35	-	39	1.41	+	38.8	1.77	+
T	32.5	0.71	-	33.5	2.12	+	34	1.41	+

S = Simplex
H = Hooke-Jeeves
B = Box-Wilson
D = Std. deviation
+ = Yes
- = No

M = Mean (S,H + B) methods
N = ID. H,B
P = ID. S.B
T = ID. S,H

U = Values within the range $N \pm 2 D_2$
Y = ID. $P \pm 2 D_3$
Z = ID. $T \pm 2 D_4$
W = ID. $M \pm 2 D_1$

REFERENCES

(1)   Kaiser R.E. and Oechich E.; "Optimization in H.P.L.C.".
      Hüthig Verlag, Heidelberg 1981

(2)   Jandera P.and Churáček J.; "Advances in Chromatography".
      vol.19, pp. 150-151 and 209-213. Edited by J.C. Giddings
      J., Grushka E. Cazes J. and Brown P.R.; Marcel Dekker
      Inc., New York 1981

(3)   Wegsheider W., Lankmayr E.P. and Budna K.W.; Chromato-
      graphia 15, (1982) 498

(4)   Rafel J. and Lema J.M.; Afinidad, in press.

(5)   Rafel J.; Poster Communication VIIth International
      Symposium on Column Liquid Chromatography. Baden-Baden,1983

(6)   Akhnazarova S. and Kafarov V.; "Experiment Optimization
      in Chemistry and Chemical Engineering". pp. 451-458. Mir
      Publ. Moscow 1982

(7)   Box G.E.P., Hunter W.G. and Hunter S.; "Statistics for
      experimenters an introduction to design, data analysis
      and Model Building". Wiley, New York 1978.

(8)   Deming S.N. and Kong R.L.; J. Chromatogr. 217, 421-434
      (1981)

(9)   Spendley W., Hext, G., R. and Himsworth F.R.; Techno-
      metrics 4, 441 (1962)

(10) Box GEP and Wilson K.B., J. Royal Stat. Soc. 13, Sec B,
      1 (1951)

(11) Rafel J., Andreu P., Mans V. and Lema J.M., Ingeniería
      Química 166 (1983) 65-71

(12) Van der Wal, Sj. and Huber J.F.K.; J. of Chromatography
      251 (1982) 290

(13) Snyder LR. and Kirkland J.J.; "Introduction to Modern
      Liquid Chromatography". 2nd Edition - John Wiley, New York
      1979

MICROCOMPUTER ANALYSIS OF HPLC REVERSED-PHASE RETENTION BEHAVIOR OF SEVERAL PHARMACEUTICAL AGENTS DEPENDING ON THE COMPOSITION OF THE ELUENT

K. VALKÓ

Institute of Enzymology, Biological Research Center,
Hungarian Academy of Sciences, P.O.Box 7, Budapest,
1502 Hungary

INTRODUCTION

Hydrophobic properties of new biologically active agents, especially the logarithm of the partition coefficients in the 1-octanol/water system /log $K_{o/w}$/ are widely used in Quantitative Structure-Activity Relationship /QSAR/ investigations[1,2,3]. The traditional shake-flask method for determining log $K_{o/w}$ has many disadvantages, such as relatively large amount of the required material changes in the partition process due to impurities, tedious and time-consuming method required for measuring the concentration in both phases. Therefore, this method is more often replaced by different chromatotographic methods such as TLC[4,5], HPLC[6,7,8,9], and GLC[10,11].

Reversed-phase liquid chromatographic retention behavior of compounds is governed by partition processes between the apolar stationary and the polar mobile phases[12,6] according to eq. 1.

$$\log k' = \log K_{ch} + \log V_s/V_m \qquad /1/$$

where k' is the capacity ratio, K is the partition coefficient of the compounds in the given chromatographic partition system, and V_s/V_m is the ratio of the volumes of the mobile and stationary phases.

The alternative chromatographic methods, other than the shake-flask method are tested by correlation analysis of log K_{ch}

values obtained by the two different methods. If only relative hydrophobicity parameters are needed for a compound series, the log k' values obtained at given HPLC conditions can be directly used for QSAR investigations. Good correlations between log $K_{o/w}$ and the log k' values for different types of compounds were obtained when 1-octanol was used as the stationary phase and water saturated with 1-octanol as the eluent[8,9]. When C-18 reversed-phase column was used good correlations between log $K_{o/w}$ and log k' values were obtained only for closely related compound series[6,7]. This method seems to be more general from a theoretical point of view, although it has practical disadvantages. For example, the chromatographic conditions cannot be optimized for a wide series of compounds and accurate reproduction of the column coated with 1-octanol seems to be difficult.

In this work we tried to develop a general relationship between log K and the HPLC retention data and to optimize the chromatographic conditions with the help of microcomputer analysis of the retention data obtained by reversed-phase HPLC of different drugs, as a function of the eluent composition.

THEORETICAL CONSIDERATIONS

According to eq. 1. the log k' values have a linear relationship to the log $/K_{chr}/$ values, where K_{chr} refers to the value of the chromatographic partition system. Collander[13] has found a linear relationship between the log K values measured in two different solvent systems,

$$\log K_{2/1} = a_1 \log K_{3/1} + b_1 \qquad\qquad /2/$$

where the subscript 1 indicates water, while subscripts 2 and 3 indicate two different apolar solvents in the partition system; a_1 and b_1 are constants obtained by the least squares method. Leo[14] showed the limitations of the validity of eq. 2, namely that the linear relationship between log $K_{2/1}$ and log $K_{3/1}$ exists only when solvents 2 and 3 have a similar character, or when the compounds considered are similar in structure. Similar limitations can be assumed for the validity of eq. 3, where $\log K_{o/w}$ and $\log K_{chr}$ were used in eq. 2:

$$\log K_{o/w} = a_2 \log K_{chr} + b_2 \qquad /3/$$

Since the HPLC chromatographic partition system cannot be kept constant for a wide variety of compounds, the problem of standardization of $\log K_{chr}$ also arises. For this reason the experimental linear relationship[15,16] between the $\log k'$ values and the percentage of the organic phase in the eluent /OP%/ for a given compound was taken into consideration in eq. 4:

$$\log k' = a_3 \; OP\% + b_3 \qquad /4/$$

The linearity of eq. 4 exists only in a given range of $OP\%$[17]; therefore, this relationship has to be tested experimentally. On the basis of eq. 4, knowing a_3 and b_3 from the experimental data by least squares analysis, the $\log k'$ values of different compounds can be extrapolated to the same $OP\%$, thus a standard value of $\log K_{chr}$ can be obtained. Log k'_o values obtained by extrapolation of $\log k'$ to the zero percentage of the organic phase were used in eq. 3 /$\log k'_o = b_3$ according to eq. 4/. A statistically significant relationship between $\log K_{o/w}$ and $\log k'_o$, cannot be obtained considering Leo's statements[14], because 1-octanol essentially differs from the chemically-bonded reversed phases in the column due to its OH function. Therefore, a characteristic value expressing the sensitivity of compounds to changes in the polarity of the partition system was sought which can be easily measured. Since the physical meaning of a_3 in eq. 4 is the change in $\log k'$ caused by a 1% addition of the organic phase in the eluent /'slope'/, this parameter is assumed to be the best for characterizing the special interactions between the solute and the solvent. Thus, the following relationship was developed and tested by experimental data:

$$\log K_{o/w} = a_4 \log k'_o + b_4 \; '\text{slope}' + c_4 \qquad /5/$$

This equation is considered to be a generalization of the relationship betwen $\log K_{o/w}$ and the RP-HPLC retention data. When closely related series of compounds were analysed[18] by eq. 4, the straight lines for each compound were parallel, that is to say the 'slope' can be considered constant in eq. 5. This is the reason why McCall[6] and Carlson et al.[7] found good linear relationships between $\log K_{o/w}$ and $\log k'$ without considering 'slope'.

MATERIALS AND METHODS

Compounds for which 1-octanol/water partition coefficients were available from the compilation of Hansch and Leo[19] were chosen as a model for testing the validity of eq. 5. Compounds were obtained from Semmelweis Medical University Budapest. Most of them are listed in the Sixth Edition of the Hungarian Pharmacopea. The World Health Organization (WHO) names of compounds and corresponding $\log K_{o/w}$ values are listed in Table I. In this work ionic compounds (salts) were not chosen in order to avoid using ion pairs during the chromatographic measurements because ion-pair formation can influence the partition processes of the compounds. The experimental conditions of the HPLC analyses are given in Table II. The concentration of acetonitrile in the eluent was increased by 5% steps from 5% to 90%. Log k' values of all compounds were determined at three to five different acetonitrile concentrations in order to test the validity of eq. 4 and to obtain a_3 ('slope') and b_3 ($\log k'_o$). The change in the pH value of the eluent from 4.6 to 2, in the case of acidic compounds, was necessary in order to obtain symmetric peaks. The effect of change in the pH is to be reflected in the value of the 'slope', thus, pH changes can be properly taken into consideration in eq. 5. All calculations including the treatment of the measured retention data and least squares calculations were carried out on an Apple II+ microcomputer using BASIC program written by us. Log k' values of all compounds measured at different acetonitrile concentrations are stored in data files, which can be continuously enlarged and easily used in correlation analysis.

742

Table I   The names of the compounds investigated and their
         log $K_{o/w}$ values[19]

Compound	log $K_{o/w}$
Resorcin	0.80
Sulfadimidin	0.32
Sulfamethoxypyridazin	0.40
Barbital	0.65
Phenobarbital	1.42
Chloramphenicol	1.14
Salicylamid	1.28
Phenacetin	1.58
Vanillin	1.37
Benzaldehyde	1.45
Acetanilid	1.16
Nicotinamid	-0.57
Benzoic acid	1.87
Salicylic acid	2.25
Acetyl salicylic acid	1.23
Coffein	-0.07
Hydrochlorothiazid	-0.07
Cortexolone	2.46
Cortisone	1.64
Dexamethasone	1.99
11-Deoxycorticosterone	2.88
Progesterone	3.87

Table II   Experimental conditions of the HPLC measurements

Column[x]: RP-18 LiChrosorb, 250 nm, ID 4.6, dp 10 μ

Injector[xx]: Rheodyne Model 7010 Sample Injection Valve

Detector[x]: ISCO Model 226 Absorbance Monitor

Detection: 254 nm

Pump: Labormim Liquopump model 312,

Integrator: Chinoin Digint model, 24

Recorder: Endin model 621.01

Temperature: $22^{\circ}C \pm 2^{\circ}C$

Eluent: 5-90% acetonitrile /Reanal, extrapurified for
   chromatogr./ and 0.05 M KH_2PO_4 buffer /pH = 4.6/[xx]

Dead time determination: $NaNO_3$

Calculations[xxx]: Apple II+ microcomputer

---

[x]   Instruments were kindly provided by Chromatronix, Inc.

[xx]   For the measurements of acidic compounds 1-2 drops of
       H_3PO_4 were added to the eluent /pH = 2/ in order to
       get symmetrical peaks.

[xxx]   The computer was kindly provided by Dime's Group Inc.

Table III Coefficients of the linear relationhips between
           log k′ and OP%

Compound	'Slope'	log k′$_o$	r	OP%
Resorcin	-0.0150	0.259	0.990	17.27
Sulfadimidin	-0.0280	0.854	0.997	30.57
Sulfamethoxypyridazin	-0.0285	0.892	0.990	31.30
Barbital	-0.0402	1.063	0.981	26.44
Phenobarbital	-0.0319	1.341	0.999	42.04
Chloramphenicol	-0.0414	1.625	0.997	39.25
Salicylamid	-0.0255	0.871	0.984	34.16
Phenacetin	-0.0226	1.002	0.981	44.37
Vanillin	-0.0244	0.866	0.999	35.49
Benzaldehyde	-0.0303	1.575	0.999	51.98
Acetanilid	-0.0270	1.021	0.991	37.81
Nicotinamid	-0.0382	0.251	0.941	6.57
Benzoic acid	-0.0284	1.252	0.987	44.08
Salicylic acid	-0.0301	1.425	0.988	47.34
Acetyl salicylic acid	-0.0272	1.077	0.974	39.60
Coffein	-0.0299	0.552	0.979	18.46
Hydrochlorothiazid	-0.0456	0.887	0.912	19.45
Cortexolone	-0.0166	0.997	0.965	60.06
Cortisone	-0.0104	0.241	0.946	23.17
Dexamethasone	-0.0168	0.821	0.996	48.87
11-Deoxycorticosterone	-0.008	0.585	0.999	73.13
Progesterone	-0.0143	1.410	0.982	98.60

r is the correlation coefficient of the straight line
OP%$_o$ is the percentage of acetonitrile at which log k′ = 0.

RESULTS AND DISCUSSIONS

Table III lists the parameters of eq. 4 for all 22 compounds. The straight lines obtained can be seen in Fig. 1 and
they are considered statistically significant within the measured range of OP%. Plotting the log k'_o values against the
log $K_{o/w}$ values in Fig. 2, it can be easily seen that in this

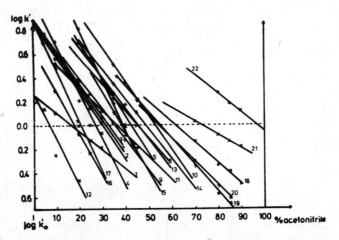

Fig. 1. The relationship between log k′ values and the concentration of acetonitrile in the eluent (OP%) for the
22 model compounds

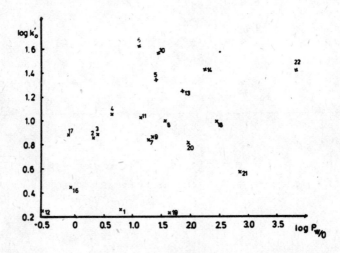

Fig. 2. The plot of log $K_{o/w}$ against log k′o

746

case the Collander-type relationship /eqs. 2 and 3/ is not valid. By calculating the coefficients of equation 5, eq. 6 was obtained:

$$\log K_{o/w} = 92.55 \text{ 'slope'} + 1.77 \log k'_{o} + 2.08 \qquad /6/$$

$$n = 22 \qquad r = 0.926 \qquad s = 0.408 \qquad F = 56.9$$

where

n is the number of compounds,

r is the multiple correlation coefficient,

s is the residual error,

F is the F-test value.

The difference between the measured and the calculated $\log K_{o/w}$ values never exceeded 0.7, consequently none of the data points can be considered as an outlier.

The correlation matrix of equation variables can be seen in Table IV. 'slope' and $\log k'_{o}$ have a low correlation coefficient. A significant improvement in the quality of statistics was obtained by introducing the second independent variable into the equation.

Table IV Correlation matrix of variables used in eq. 6

	$\log K_{o/w}$	$\log k'_{o}$	'slope'
$\log K_{o/w}$	1.000	0.387	0.658
$\log k'_{o}$	0.387	1.000	-0.342
'slope'	0.658	-0.342	1.000

As a conclusion, it can be stated that the experimental data supported the theoretical considerations. We also intend to use this relationship for predicting the eluent composition for compounds with known $\log P_{o/w}$ values. For this reason, on the basis of the $\log k'$ vs. OP% relationship, the OP% values were calculated for $\log k'$ equaling zero for each compound /OP%$_{o}$/. /Table III/. This retention time is equal to a value

twice the dead time at given HPLC conditions. Using eq. 4 with
log k' = 0 eq. 7 is obtained:

$$OP\%_o = - \frac{\log k'_o}{'slope'}$$ /7/

It can be seen that the $OP\%_o$ values involve both independent
variables of eq. 5. Plotting $OP\%_o$ against log $K_{o/w}$ in Fig. 3
a statistically significant linear relationship was obtained
/eq. 8/:

$$\log K_{o/w} = 0.047 \ OP\%_o - 0.536$$ /8/

$$n = 22 \qquad r = 0.918 \qquad s = 0.407$$

This equation makes it possible to predict $OP\%_o$ values on
the basis of log $K_{o/w}$ values of given compounds at which log k'
equals zero. Similar equations can be set up using OP% values
for predicting arbitrary log k' values of individual compounds.

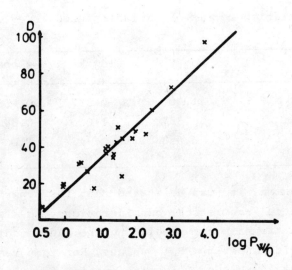

Fig. 3. The relationship between $OP\%_o$ and log $K_{o/w}$ values
on the basis of eq. 8

CONCLUSIONS

A general relationship between the logarithm of the partition coefficients in the 1-octanol/water system /log $K_{o/w}$/ and the RP-HPLC retention data was described using eqs. 5 and 8. These can be used for predicting log $K_{o/w}$ values of compounds on the basis of their reversed-phase retention behavior. The relationship /eq. 8/ is also applicable for designing proper eluent compositions for compounds having known log $K_{o/w}$ values. All calculations and data handling have been computerized.

ACKNOWLEDGEMENT

The Apple II+ microcomputer was kindly provided by Dime's Group Inc. The injection valve, the absorbance monitor, columns and other chromatographic supplies were supplied by Chromatronix Inc.

The author gratefully thanks Drs. F. Bartha and Gy. Mátrai for their valuable contributions in the field of mathematical statistics and computer programming and Mrs. J. Báthi for her skilled technical assistance.

REFERENCES

1. C. Hansch and T. Fujity; J. Am. Chem. Soc., 86 /1964/, 1616.

2. C. Hansch and W.J. Dunn; J. Pharm. Sci., 61 /1972/, 1.

3. J. K. Seydel and K.J. Schaper; Chemische Struktur und Biologische Aktivität von Wirkstoffen, Verlag Chemie, Weinheim, 1979.

4. C.B.C. Boyce and B.V. Milborrow; Nature /London/, 208 /1965/, 537.

5. E. Tomlinson; J. Chromatogr. 113 /1975/, 1.

6. J.M. McCall; J. Med. Chem., 18 /1975/, 6, 549.

7. R.M. Carlson, R.E. Carlson and H.L. Koppermann; J. Chromatogr., 107 /1975/, 219.

8. M.S. Mirrlees, S.J. Moulton, C.T. Murphy and P.J. Taylor; J. Med. Chem., 19 /1976/, 5, 615.

9. S.H. Unger, J.R. Cook and J.S. Hollenberg; J. Pharm. Sci., 67 /1978/ 10, 1364.

10. O. Papp, K. Valkó, Gy. Szász, I. Hermecz, J. Vámos, K. Hankó-Novák and Zs. Ignáth-Halász; J. Chromatogr., 252 /1982/ 67.

11. K. Valkó and A. Lopata; J. Chromatogr. 252 /1982/, 77.

12. W.J. Haggerty, Jr., and E.A. Murrill; Res. Dev., 25 August /1974/ 30.

13. R. Collander; Acta Chem. Scand., 5 /1951/, 774.

14. A.J. Leo in 'Biological Correlations - The Hansch Approach' Advances in Chemistry Series, No. 114, Ed. R.F. Gould, American Chemical Society Washington, D.C. 1972. p. 51.

15. Cs. Horváth, W. Melander and I. Molnár; J. Chromatogr., 125 /1976/, 129.

16. F.J. Yang; J. Chromatogr. Sci., 20 /1982/, 241.

17. C.M. Riley, E. Tomlinson and T.M. Jefferies; J. Chromatogr., 185 /1979/, 197.

18. M.L. Cotton and G.R.B. Down; J. Chromatogr., 259 /1983/ 17.

19. C. Hansch and A. Leo, Substituent Constants for Correlation Analysis in Chemistry and Biology, Wiley, New York, 1979.

SUBSTITUENT EFFECTS IN RPTLC

É. JÁNOS, B. BORDÁS, T. CSERHÁTI and G. SIMON*

Research Institute for Plant Protection, Hungarian Academy
of Sciences, Budapest, Herman Ottó u. 15. 1022 - Hungary
*Institute of Pathophysiology, Semmelweis University
of Medicine, Budapest, Hungary

SUMMARY

R_M values of 36 quaternary ammonium steroids were determined by reversed-phase thin-layer chromatography. The R_M
values of these compounds - deviating from the general role -
do not change linearly with the organic solvent ratio of the
eluent. Increasing the quantity of organic solvent in the mobile phase the R_M values decrease /as usual/ but above a certain concentration /about 80 percent organic phase content/
they increase again. We obtained similar results with several
solvent systems and several types of reversed-phase plates.
Stepwise regression analysis was applied to find the responsible substituent groups.

INTRODUCTION

The Quantitative Structure - Activity Relationship /QSAR/
methods are useful and wide-spread in designing new and biologically effective molecules. For these methods we need data
on molecular lipophilicity, which can be measured by several
ways. First of all a classic way is to measure the partition
coefficient in octanol-water system /1, 2/. A more simple but
exact and quick method is the use of reversed-phase thin-layer
chromatography /RPTLC/ /3, 4/. As the high-pressure liquid chromatography /HPLC/ becomes more and more generally applied, a
new and convenient method is available for the determination of
lipophilicity by reversed-phase HPLC /5, 6/. A recent attempt
to determine lipophilicity is by gas chromatography /7, 8/.

The lipophilicity of the steroid derivatives was measured
by RPTLC. By this way we determined the R_f values of these molecules using different solvent systems and calculated the R_M
values according to the equation:

$$R_M = \log \left| \frac{1}{R_f} - 1 \right|$$

The measured R_M values did not follow the general rule. They
began to increase above a certain organic solvent concentration. Multivariate mathematical methods were used for the establishment of the responsible substituent groups.

Experimental

The substituent groups of the steroid molecules are shown
in Fig. 1.

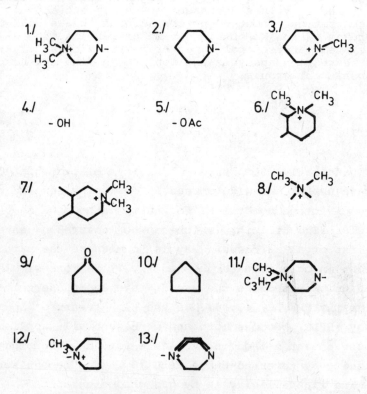

Fig. 1   The substituents

We used precoated Silicagel plates impregnated with 5 % paraffin oil in n-hexane.

The eluent systems were
1. dioxane-water   70-30
2. dioxane-water   80-20
3. dioxane-water   90-10

For the Overpressured Thin layer Chromatographic /OPTLC/ separations a Chrompres 10 chamber /Labor MIM, Hungary/ was used /9/.

The spots were visualised in J_2 chamber.

For the analysis of our data set we have applied a stepwise regression analysis.

RESULTS AND DISCUSSION

R_M values measured in different eluent systems are listed in Table I.

Analysing these data a peculiar phenomenon could be observed. Usually the R_m values become smaller with increasing concentration of the organic constituent of the eluent system. However, this was not true in our case. In the applied eluent systems the molecules investigated moved faster when the quantity of the organic component was increased. This agrees with the general rule. However, having reached a certain concentration /about 80 percent of the organic constituent/ their movement slowed down again. We tried to find an explanation for this anomaly. We assumed that this was the result of the imperfect paraffin coating of silica gel: the increasing quantity of the organic solvent present in the mobile phase may have ruined the paraffin coating and therefore, the plates could not be considered as a reversed-phase anymore. In order to eliminate this possibility we repeated the development on silanized silica gel plates which are not sensitive to organic solvents. However, in this case the result was the same, we could observe the same phenomenon. This finding may be due to the specific structure of these compounds containing extremely lipophilic and hydrophilic parts in one molecule. Earlier we observed si-

Table I. R_M values of the steroid derivatives in different eluent systems.

number	eluent system		
	1.	2.	3.
1.	-0.53	-0.30	0.86
2.	-0.92	-1.20	-0.47
3.	-0.74	-0.54	0.50
4.	-0.51	-0.46	0.50
5.	-0.60	-0.54	0.44
6.	-0.63	-0.26	-0.25
7.	-0.23	-0.15	0.73
8.	-0.17	0.15	1.11
9.	-0.92	-1.24	-0.39
10.	-0.26	-0.17	0.82
11.	-0.63	-1.24	-0.44
12.	-0.92	-1.20	-0.53
13.	-0.28	-0.12	0.77
14.	-0.92	-1.24	0.20
15.	-0.17	0.15	1.10
16.	-0.63	-0.47	0.44
17.	0.10	0.35	1.43
18.	-0.15	0.28	1.30
19.	-0.50	-0.01	1.11
20.	-0.39	-0.07	0.97
21.	-0.03	0.65	1.36
22.	-0.20	0.44	1.36
23.	-0.86	-0.62	0.47
24.	-0.14	0.11	1.24
25.	-0.07	0.20	1.30
26.	-0.59	-0.36	0.77
27.	-0.32	-0.10	1.03
28.	-0.09	0.25	1.43
29.	-0.05	0.29	1.30
30.	-0.14	0.16	1.30
31.	-0.12	0.06	1.19
32.	-0.07	0.32	1.51
33.	-0.09	0.34	1.43
34.	-0.12	0.33	1.30
35.	-0.01	0.37	1.36
36.	-0.05	0.41	1.36

milar anomaly in the case of other compounds, namely in the case of morphine derivatives.

In the present study we tried to establish which substituent groups are responsible for this anomaly, using stepwise regression analysis. The results are listed in Table II.

In Table II the a coefficients represent contribution of the skeleton to the R_M values while the b coefficients show the contribution of the substituent to R_M. The higher the t value the more significant the contribution of the substituent group.

According to the calculation, in dioxane-water 70:30 two substituent groups influence the R_M values. One is the hydroxyl group /substituent 4/, while the other is substituent 9. The hydroxyl group decreases the R_M value, while the other substituents containing quaternary ammonium part increase it.

In the following eluent system /dioxane-water 80:20/ -OH, -OAc, substituent 2 and substituents containing quaternary ammonium parts /1, 6, 7/ significantly influence the R_M value.

What are the significantly influencing substituents in the eluent system dioxane-water 9:1 where the anomaly was observed and where the R_M values dramatically increased? Two substituents were found /6 and 8/ which significantly influenced the R_M value, both containing quaternary ammonium parts.

The regression analysis supports our hypothesis, namely that the quaternary ammonium compounds behave irregularly in chromatographic systems.

Cs. Horváth observed the same phenomenon in the case of crown ethers /10/. He established that in HPTLC the surface silanols play a major role when large molecules having polar functions are chromatographed using water-lean binary eluents in reversed-phase chromatography. According to this model the magnitude of eluite retention can be determined not only by solvophobic /hydrophobic/ interactions but also by reversible binding involving polar silanolic sites on the surface. The term silanophilic interactions was used by him to denote binding mechanism with the participation of silanol groups. The observed "irregular" retention behaviour of quaternary amino steroids in RPTLC can be explained by a similar retention mechanism: the

Table II Results of stepwise regression analysis

A./ Independent variables selected:
substituents 5, 4, 2, 7, 8 and 6.
Dependent variable: R_M values measured in
dioxane - water 70 : 30

$t_{95\%} = 2.05$ $F_{99\%} = 3.59$
a coeff.:   -399.532

b coeff.:	sb values	t test	path coeff.
S5 -308.999	271.662	1.137	-0.172
S4 -363.981	157.231	2.315	-0.340
S2 -521.465	263.510	1.979	-0.290
S7 -327.967	191.858	1.709	-0.254
S8 294.531	143.160	2.057	0.312
S6 210.531	109.532	1.922	0.304

R = 0.677   s = 255.454   F = 3.665

B./ Independent variables selected:
substituents 4, 1, 5, 1', 2, 7, 8 and 6.
Dependent variable: R_M values measured in
dioxane - water 80 : 20

$t_{95\%} = 2.06$ $t_{99\%} = 2.79$ $F_{99.9\%} = 4.99$
a coeff.: -208.663

b coeff.:	sb values	t test	path coeff.
S4 -968.062	271.191	3.570	-0.517
S1 387.748	258.884	1.498	0.207
S5 -435.797	208.468	2.091	-0.312
S1' 466.584	215.398	2.166	0.334
S2-1035.330	393.993	2.628	-0.330
S7 -737.212	315.632	2.336	-0.327
S8 332.975	228.017	1.460	0.202
S6 371.994	162.471	2.290	0.3078

R = 0.802   s = 377.484   F = 5.389

C./ Independent variables selected:
substituents 2, 12, 13, 2', 8 and 6
Dependent variable: R_M values measured in
dioxane - water 90 : 10

$t_{99\%} = 2.79$ $F_{99\%} = 3.59$
a coeff.: 483.975

b coeff.:	sb values	t test	path coeff.
S2 450.523	372.062	1.211	0.167
S12-277.650	275.179	1.009	-0.168
S13-926.960	511.091	1.814	-0.249
S2'-685.968	511.093	1.342	-0.184
S8 1080.010	335.459	3.220	0.553
S6 753.506	211.108	3.569	0.526

R = 0.724   s = 495.554   F = 4.772

retention is caused not only by the usual solyophobic interactions but also by "silanophilic" interactions between the eluite and the accessible silanol groups at the surface of the paraffin oil coated thin-layer plates. This hypothesis is supported by our mathematical analysis which shows that mainly the substituents containing quaternary ammonium parts /polar functions/ influence the R_M values.

REFERENCES

1. T. Fujita, J. Iwasa, C. Hansch: J. Am. Chem. Soc. 86 5175-80 /1964/

2. C. Hansch, S.M. Anderson: J. Org. Chem. 32 2583 /1967/

3. C.B.C. Boyce, B.V. Millborrow: Nature, 208 537 /1965/

4. G.L. Biagi, A.M. Barbaro and M.C. Guerra: J. Chromatogr. 41 371 /1969/

5. J.M.Mc. Call: J. Med. Chem. 18 549 /1975/

6. M.S. Mirrlees, S.J. Moulton, C.T. Murphy and P.J. Taylor: J. of Med. Chem. 19 615 /1976/

7. K. Bocek: J. of Chromatogr. 162 209-214 /1979/

8. O. Papp, G. Szász, K. Valkó: Acta Pharm. Fenn. 90 107 /1981/

9. E. Tyihák, E. Mincsovics and H. Kalász: J. Chromatogr. 174 75-81 /1979/

10. A. Nahum and Cs. Horváth: J. Chromatogr. 203 53-63 /1981/

VARIOUS TOPICS

CHROMATOGRAPHIC ANALYSIS OF PURINE AND PYRIMIDINE DERIVATIVES FROM THE TISSUES OF ANIMALS AND HUMANS IN EXPERIMENTAL AND CLINICAL PATHOLOGY

R.T. TOGUZOV, Yu. V. TIKHONOV, I.S. MEISNER, P.A. KOL'TSOV,
O.Ya. POLYKOVSKAYA and Yu. S. BUTOV

Biochemistry Department, Central Research Laboratory,
N.I. Pirogov 2nd Moscow Medical Institute,
1, Ostrovityanova Ul., 117437 Moscow, USSR
Hospital Therapy Department No. 2.
N.A. Semashko Moscow Medical Stomatologic Institute,
20/1, Delegatskaya Ul., 103473 Moscow, USSR

SUMMARY

A combination of liquid anion-exchange chromatography and high-performance liquid chromatography has been used in the determination of the pools of purine and pyrimidine derivatives in different organs of normal rats and mice and in experimental pathologic states.

The tissue and organ specificity of the qualitative and quantitative compositions of acid-soluble fractions /ASF/ of intact organs has been demonstrated.

Experimental models of regenerative hypertrophy /mouse liver regenerating after partial hepatectomy/, autolytic destruction /various periods of the dying of the organism/, and of variation dynamics of the composition of the liver ASF in the course of a day /diurnal rhythm/ have revealed a distinctively stereotyped phase-like accumulation of purines and pyrimidines in the liver of the animals.

In order to make a comparative analysis and to identify correlative connections between different components of complex mixtures of substances obtained for each of the studied models, coefficients characterizing the ASF have been developed; these coefficients permit a general evaluation of the direction of the metabolism of purine and pyrimidine derivatives in each concrete case.

These findings have promoted the use of the method of "comparative chromatograms" obtained by high-performance liquid chromatography in analyzing the ASF of erythrocytes, plasma, the mucous membrane of the stomach and of the tumour tissue in patients with carcinoma of the stomach. The diagnostic value of the method in the detection of the focus of neoplastic growth in the organism has been shown.

INTRODUCTION

Most pathologic phenomena occurring in the organs and tissues always involve destruction and repair processes which depend on a number of factors: the activity, gravity, and stage of the disease, medicamental therapy, etc. The arising pathology - both of focal and systemic character - is accompanied by concrete biochemical disturbances on the cellular metabolism level. First of all, these processes should affect the metabolism of nucleotides and their derivatives since these compounds not only ensure the biosynthesis of DNA and RNA but also perform many special functions in energy, carbohydrate, and lipid metabolism.

It is well known that purine and pyrimidine derivatives are carried by blood and that blood erythrocytes are a direct system of transport of nucleotides /1/. Therefore, the analysis for the metabolites of the purine and pyrimidine series in tissues and blood in different diseases may not only reflect the development of the pathologic process but also be a method of diagnosis and prognosis of the course of diseases.

The methods of high-performance liquid chromatography are widely used for the determination of purine and pyrimidine derivatives contained in the acid-soluble fraction /ASF/ of tissues of the organism /2, 3/. However, a number of difficulties /stringent requirements for the preliminary treatment and purification of the sample, insufficient selectivity and efficiency of chromatographic columns for the analysis of multicomponent mixtures of substances, etc./ limit this type of chromatographic separation mainly to studies of biological fluids /blood cells and plasma, lymph, urine/ and cell cultures with the analysis of the composition of nucleosides and nitrogenous bases /4, 5/. Only a few studies dedicated to the chromatographic separation of nucleotides present in animal tissues are known /6, 7/. At the same time a necessity to study the whole range of the components of nucleic acids /including their metabolites and derivatives/ in normal and pathologically changed tissues and organs of animals and humans arises not infrequently in experimental work and in the clinic.

For this purpose we have developed a combined chromatographic method for the analysis of the components present in
the acid-soluble fraction of animal organs and tissues with
the use of different chromatographic systems, packing materials
and elution modes.

MATERIALS AND METHODS

Purine and pyrimidine nucleotides, nucleosides, and N-ba-
ses were purchased from Sigma /St. Louis, MO, USA/. $NH_4H_2PO_4$,
KOH, NaCl, and $HClO_4$ were obtained from Merck /Darmstadt, FRG/.
Water was purified with the Elgastat Water Purification Systems
/The Elga Group, UK/.

ANIMALS

For all experiments, 180 g male Wistar rats and 18-20 g
male Fl C57/Bl6 x CBA mice were used. For regenerating liver
experiments, 18-20 g male Fl C57/Bl6 x CBA mice were used, and
partial hepatectomy was performed by the standard procedure
/8/. Sham-operated Fl mice were taken as controls. The animals
were kept under usual vivarium conditions. One day before the
experiment, the animals were kept in separate cages and received only water.

PREPARATION OF THE ACID-SOLUBLE FRACTION

Animals were lightly anesthetized with ether; the organs
to be studied were exposed, weighed, and placed without delay
in a porcelain mortar with liquid nitrogen. The tissue was
ground with a porcelain pestle to a powder-like state, and
cold 0.6 N perchloric acid /1 · 5 w/v/ was added. The mixture
was homogenized in a homogenizer with a Teflon pestle /10 - 15
strokes/. The homogenate was centrifuged for 5 min at 4°C in a
K-23 /GDR/ centrifuge /5000 rpm/. The supernatant was decanted

and the sediment treated with 0.3 N perchloric acid /half of the initial volume/ and centrifuged under the same conditions. The supernatants were joined and the pH of the liquid was raised to 8.1 with concentrated KOH solution. The precipitate /$KClO_4$/ was separated by a brief centrifugation and the solution lyophilized to dry residue in a New Brunswick Scientific dryer. The sample obtained in this way was redissolved in 5 ml of deionized water and filtered through a GF/F Whatman filter.

Patients with carcinoma of the stomach and normal controls. Five individuals ranging from 55 to 70 years with established diagnosis of the IVth stage carcinoma of the stomach were studied. Ten healthy adult volunteers served as normal controls. The biopsy material of the stomach tumour tissue and of the adjacent and remote areas of the mucous membrane was obtained by an Olympus /Japan/ fibrogastroscope. Venous blood samples were taken from the same patients and also from individuals with gastritis and normals.

Tissue and blood sample treatment. The obtained biopsy samples /5-10 mg/ were frozen in liquid nitrogen, the frozen tissue was pulverized in a liquid nitrogen-cooled mortar, and the powder was extracted in 1 ml of aqueous $HClO_4$ /0.6 N/. The extracts were centrifuged in order to remove the precipitated protein and then neutralized with KOH. After centrifugation to remove $KClO_4$, the pH of the samples was brought to 3.5 by the addition of HCl. Aliquots of these extracts were analyzed by anion-exchange chromatography /HPLC/.

Plasma and erythrocytes were prepared by adding 10 ml of blood to a tube treated with sodium citrate and immediately centrifuging for 10 min at 2000 x g /4°C/. Acid-soluble fractions of plasma and erythrocytes were also obtained by double $HClO_4$ extraction. The first extraction was carried out by 5 volumes of 1N $HClO_4$ and the second one by a 2.5-fold volume of 0.5 N $HClO_4$. After centrifugation, the supernatant fluids were joined and their pH was brought to 3.5 with KOH. $KClO_4$ was spun down for 5 min at 10000 x g /4°C/. Aliquots of these extracts were directly injected into the HPLC column.

CHROMATOGRAPHIC ANALYSIS OF THE ACID-SOLUBLE FRACTION OF ANIMAL ORGANS

At the first stage of the study the samples were prepared by anion-exchange chromatography on a Hitachi Model 034 instrument.

Chromatographic conditions:

Column:	0.9 cm x 10 cm
Packing material:	-630, Cl^- /Hitachi/
Mobile phase:	Six communicating vessels of a Varigrad gradient mixer filled with solutions /30 ml each/ of the following compositions:

1. 10^{-3} N HCl   pH 2.95
2. 10^{-2} M NaCl in 10^{-3} N HCl   pH 2.90
3. 2.5×10^{-2} M NaCl in 10^{-3} N HCl   pH 2.85
4. 5.0×10^{-2} M NaCl in 10^{-3} N HCl   pH 2.82
5. 5.0×10^{-2} M NaCl in 10^{-2} N HCl   pH 1.88
6. 4.0×10^{-1} M NaCl in 10^{-2} N HCl   pH 1.70

Flow rate:	2 ml/min
Temperature:	$37^\circ C$
Detector:	UV, simultaneous recording at 250, 260, and 280 nm
Sample volume:	5 ml

The eluate accumulated in the collector /5 ml in each test tube/ was analyzed on a Hitachi Model 624 double-beam scanning spectrophotometer in the 300-200 nm wavelength range. The constituent compounds were identified on the basis of retention times in the chromatogram and of their absorption spectra. The concentrations of the substances were calculated by the following formula:

$$C = \frac{/D_{max} - D_{290}/ \times K}{W}$$

where C = substance concentration / μM/100 g wet weight/

D = optical density of the sample

K = coefficients calculated on the basis of calibration

curves of standard substances such as adenine /48.8/.
guanine /71.5/, uracil /74.0/, thymine /62.2/, cytosine /130.6/, xanthine /95.4/, hypoxanthine /54.7/,
and their derivatives

W =   wet weight

The method of conventional column chromatography gives good
results for appropriate purposes but features low sensitivity
/the main shortcoming/ and insufficient selectivity for the
separation of multicomponent mixtures. For this reason, at the
second stage of our studies we have performed a new series of
analyses of fractions obtained on a Hitachi Model 034 liquid
chromatograph; different methods of high-performance liquid
chromatography have been employed.

The fractions of nucleosides and nitrogenous bases of nucleic acids unretainable by anion-exchange resin were joined
together, lyophilized to dry residue, and redissolved in deionized water. After bringing the pH of the solution to 2.5,
the sample was introduced into a Perkin-Elmer Model 604 liquid chromatograph.

Chromatographic conditions:

Column:	9.4 mm x 500 mm
Packing material:	Partisil-10 SCX
Mobile phase:	0.025 M $NH_4H_2PO_4$, 0.16 M KCl, pH 2.5, isocratic mode
Flow rate:	3.2 ml/min
Temperature:	ambient
Detector:	UV, wavelength 254 nm, 0.02 AUFS

The collected data were processed on a M-1 digital integrator /Spectra-Physics/. The calibration coefficients were
calculated automatically after the introduction of known a-
mounts of standards of nucleosides and nitrogenous bases.

Mono-, di-, and triphosphates were analysed by two methods.

1. To check the chromatographic uniformity of the peaks
corresponding to certain nucleotides, the eluate from each
test tube of the collector was introduced directly into an
Analyst 7800 liquid chromatograph /LDC/.

Chromatographic conditions:

Column: 4.6 mm x 250 mm

Packing material: Partisil-5 ODS

Mobile phase: 0.02 M $NH_4H_2PO_4$, pH 3.3, isocratic mode

Flow rate: 2 ml/min

Detector: UV, wavelength 254 nm, 0.02 AUFS

Sample volume: 200 μl

The collected data were processed on a CCM-2 /LDC/ specialized computer. Calibration coefficients were also calculated automatically. The computer permitted the statistical analysis of the results.

2. The eluates were joined in zones corresponding to mono-, di-, and triphosphates of the nucleosides, lyophilized to dry residue, and redissolved in 10 ml of deionized water. After adjusting the pH of the solution to 3.3, an aliquot was introduced directly into the system of an Analyst 7800 liquid chromatograph /LDC/. The chromatographic conditions were identical with those employed in the first case.

Note: The separation of the nucleotide material on a reversed-phase column is performed with the addition of an organic modifier /methanol, acetonitrile, etc./ and also ion-pairing agents e.g., tetrabutylammonium iodide /9/ to the eluting solution. We have shown that in our case the addition of these components to the mobile phase affects by no means the results. This made it possible to use only weak solutions of ammonium phosphate.

RESULTS AND DISCUSSION

Combined chromatographic analysis

The employed method of chromatographic separation of derivatives of the purine and pyrimidine series permits the qualitative and quantitative analyses of multicomponent mixtures of the substances present in the ASF of different animal organs.

Fig. 1 presents a typical chromatogram of the ASF of a

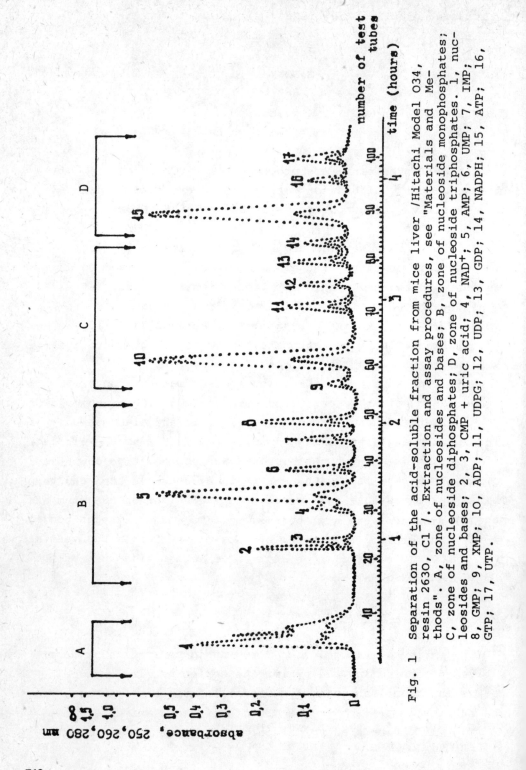

Fig. 1 Separation of the acid-soluble fraction from mice liver /Hitachi Model 034, resin 2630, Cl$^-$/. Extraction and assay procedures, see "Materials and Methods". A, zone of nucleosides and bases; B, zone of nucleoside monophosphates; C, zone of nucleoside diphosphates; D, zone of nucleoside triphosphates. 1, nucleosides and bases; 2, 3, CMP + uric acid; 4, NAD$^+$; 5, AMP; 6, UMP; 7, IMP; 8, GMP; 9, XMP; 10, ADP; 11, UDPG; 12, UDP; 13, GDP; 14, NADPH; 15, ATP; 16, GTP; 17, UTP.

mouse liver homogenate recorded on a Hitachi Model 034 liquid chromatograph. The chromatogram contains four zones /A, B, C, D/ which have been subjected to further chromatographic analyses:

A - the results of high-performance liquid chromatographic analysis of nucleosides and nitrogenous bases are given in Fig. 2;

B - Fig. 3 shows an example of the separation of the contents of four test tubes from a Hitachi Model 034 instrument corresponding to the chromatographic peak of adenosine monophosphate. It can be seen that the peak is not homogeneous and contains considerable amounts of NAD^+.

Strictly speaking, repeated high-performance chromatography of each chromatographically "pure" peak has shown that the peaks contained 2 to 4 components /10 to 20% with respect to the basic substance/ whose identification may be a subject for a separate study /deoxy forms, stereo isomers, minor bases, 2', 3', 5' isomers, etc./.

Fig. 4 presents a chromatogram of the whole zone of nucleoside monophosphates /zone B in Fig. 1/ obtained on a reversed-phase column.

Thus, the combination of different methods of liquid column chromatography and the use of various packing materials and elution modes permit the fractionation of complex multicomponent mixtures of substances contained in biological samples.

Contents of purine and pyrimidine derivatives in the acid-soluble fraction of rat organs

Our data for the contents of purine and pyrimidine derivatives in the tissues of rat liver, kidneys, spleen, skeletal muscles, lungs, brain, and testes are presented in Table I. The total content of the nucleotide material in all organs under study is variable, being the highest in the skeletal muscles, liver, and spleen and the lowest in the lungs. The total content of the ASF components in the organs is basically defined by purine derivatives /except for the spleen where the purine/pyrimidine ratio is the lowest and equals 2.3/.

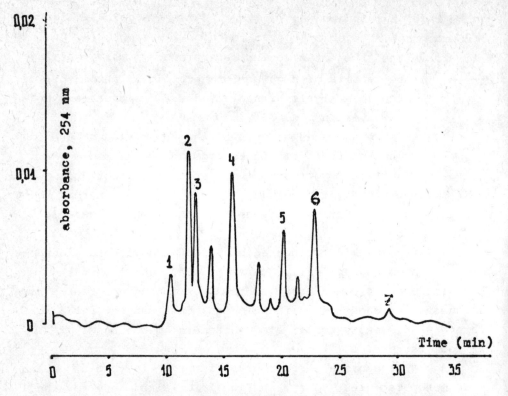

Fig.2   Isocratic HPLC separation of the fractions'
         nucleosides and bases, unretainable by anion
         exchange resin /see zone A, Fig. 1/.
         1, inosine; 2, guanosine; 3, uracil; 4, hy-
         poxanthine; 5, guanine; 6, adenosine; 7, ade-
         nine.

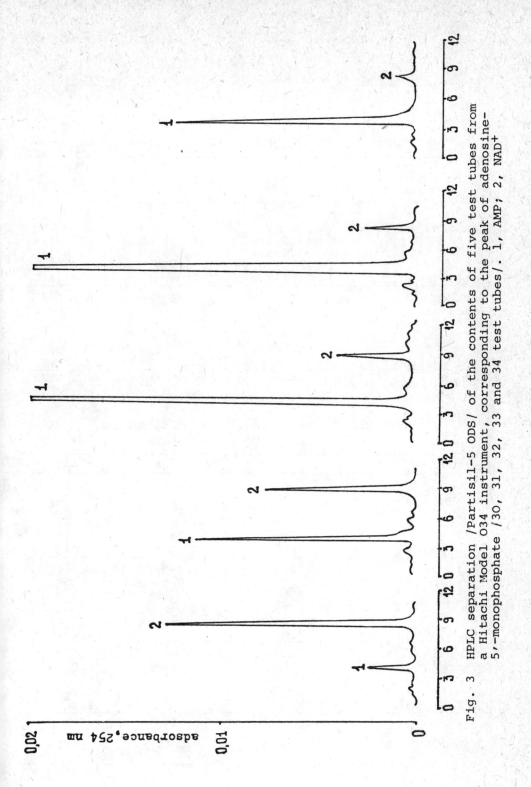

Fig. 3 HPLC separation /Partisil-5 ODS/ of the contents of five test tubes from a Hitachi Model 034 instrument, corresponding to the peak of adenosine-5'-monophosphate /30, 31, 32, 33 and 34 test tubes/. 1, AMP; 2, NAD$^+$

Table I. Purine and pyrimidine derivatives in

Extraction and assay procedures were
Data represent mean values of five

	Purine and pyrimidine		
	LIVER	KIDNEYS	SPLEEN
Adenine	19.6±3.1	7.6±0.2	17.2±2.3
Adenosine	10.6±1.7	3.2±0.1	5.1±0.6
AMP	115.5±23.2	72.4±10.5	29.4±3.1
ADP	102.8±9.9	98.3±15.9	73.8±8.3
ATP	116.6±20.5	126.8±17.8	126.8±15.1
Total adenine deriv.	365.1±32.7	308.0±26.1	252.4±17.7
Guanine	O	O	8.2±1.3
Guanosine	18.2±2.8	O	O
GMP	18.7±7.7	15.5±3.6	18.4±1.2
GDP	13.1±3.6	O	26.7±3.1
GTP	20.1±8.4	8.1±0.5	9.4±0.8
Total guanine deriv.	70.1±12.3	23.6±3.6	62.7±3.7
Hypoxanthine	9.9±1.5	12.3±3.5	21.3±3.7
Inosine	9.4±1.2	2.8±0.4	7.2±0.3
IMP	25.0±4.8	23.7±4.1	5.6±0.1
Total hypoxant. deriv.	44.3±5.25	38.8±5.5	34.1±3.5
XMP	4.4±0.9	7.9±0.8	8.4±0.5
Total PURINES	483.9±35.3	378.3±26.9	357.6±18.4
Uric acid	6.2±1.4	12.4±1.6	O
Thymine	O	O	4.2±0.2
Thymidine	7.5±1.2	2.4±0.1	O
TMP	O	O	22.3±3.8
TDP	O	O	5.2±0.1
Total thymine deriv.	7.5±1.2	2.4±0.1	39.0±3.8
Uracil	21.4±2.6	5.9±0.7	16.7±1.9
Uridine	O	3.8±0.3	O
UMP	22.5±1.2	17.3±4.1	16.3±1.2
UDP	21.4±5.1	22.6±5.5	26.5±3.7
UTP	15.7±5.0	8.5±0.9	8.1±0.5
Total uracil deriv.	81.0±7.7	58.1±7.0	67.6±4.4
Cytosine	O	O	3.5±0.2
Cytidine	O	O	6.8±0.5
CMP	7.5±0.8	9.3±1.3	29.1±2.9
CDP	10.8±1.7	O	6.2±0.9
Total cytosine deriv.	18.3±1.9	9.3±1.3	45.6±3.1
Total PYRIMIDINES	106.8±8.0	69.8±7.1	152.2±6.6
TOTAL COMPONENTS	596.9±36.2	460.5±27.8	509.8±19.6

the acid-soluble fractions of the rat organs

described in "Materials and Methods".
experiments ± S.E.

contents / µmol/100 g wet wt/			
MUSCULAR TISSUE	LUNGS	BRAIN	TESTES
27,5±5.3	20.3±3.2	4.3±0.3	1.6±0.3
16.7±2.8	6.8±0.7	O	2.7±0.3
86.0±8.9	69.4±12.6	23.7±3.2	83.8±8.2
137.3±19.6	14.8±7.9	58.8±6.1	104.4±9.2
196.5±21.7	17.3±2.1	149.7±13.6	103.5±8.1
414.0±31.2	158.7±15.4	236.6±15.3	295.6±14.8
O	O	O	0.5±0.2
Q	Q	Q	Q
18.3±1.0	12.1±2.5	26.4±3.8	15.4±2.2
9.4±0.5	9.8±1.2	20.2±3.6	19.3±1.5
2.7±0.3	4.7±0.2	7.8±0.5	4.8±0.9
30.4±1.2	27.1±4.2	54.2±5.3	40.0±2.8
14.2±0.8	24.5±3.8	O	O
5.2±0.2	62.7±5.3	Q	Q
53.6±7.1	9.6±0.8	16.4±1.6	6.3±0.5
73.0±7.3	96.8±9.8	16.4±1.6	6.3±0.5
24.8±1.6	0.9±0.1	O	O
542.2±32.1	238.5±18.7	307.2±16.3	341.0±15.1
31.2±7.0	12.3±2.3	O	O
Q	O	Q	Q
2.3±0.4	O	5.6±1.0	2.9±0.7
O	O	O	O
Q	O	Q	Q
2.3±0.4	O	5.6±1.0	2.9±0.7
11.5±2.2	26.0±3.5	1.3±0.1	2.0±0.3
2.6±0.1	Q	Q	Q
12.7±1.6	3.2±0.2	16.2±2.2	16.3±1.2
8.7±0.5	5.4±0.3	20.4±3.1	34.0±2.8
13.4±2.3	1.7±0.1	3.2±0.1	Q
48.9±3.6	36.3±3.5	41.1±3.8	52.3±3.1
O	O	O	Q
Q	Q	Q	2.6±0.3
38.7±4.2	32.7±5.5	9.7±1.6	15.9±2.1
15.3±0.9	Q	Q	Q
54.0±4.3	32.7±5.5	9.7±1.6	18.5±2.1
105.2±5.6	69.0±6.5	56.4±4.2	73.7±3.8
678.6±33.3	364.8±19.9	363.6±16.8	415.6±15.6

The chromatographic analysis of the ASF of these organs has revealed individual distinction in the compositions. Thus, guanine, thymine, cytosine, cytidine, TMP, TDP, CTP, NADPH were not found in the ASF of rat liver. On the other hand /unlike other organs/, guanosine was identified /18.2 μM/100 g/. No guanine, guanosine, thymine, cytosine, cytidine, TMP, CDP, CTP were determined in the kidneys. The spleen did not contain guanosine, thymidine, uridine, uric acid but revealed the presence of guanine, TMP, TDP, cytosine, cytidine. These substances were not found in other organs of normal animals. No guanine, guanosine, thymine, cytosine, cytidine, CDP, CTP were found in the muscles and no guanine, guanosine, thymidilates, uridine, cytosine, cytidine, CDP in the lungs. Adenosine, guanine, guanosine, hypoxanthine, inosine, thymine, uridine, cytosine, cytidine, XMP, TMP, TDP, CDP, CTP, and uric acid were not detected in the brain tissue.

It is to be emphasized that the zero concentration of any component of the ASF of an organ does not exclude its presence, it only means amounts below the threshold sensitivity of a given method.

The organs under study strongly differ in the percentages of purine and pyrimidine derivatives. Thus, in the skeletal muscles guanine derivatives constitute 3.8% and in the spleen 11.7% of the total amount; the content of cytidine derivatives in the kidneys is 1.3% and in the lungs 8.7%. The concentration of thymidine derivatives in the lungs is nil and in the spleen 7.3%. The percentage of adenine and uridine derivatives differ to a smaller degree: 42.3% /lungs/ and 61.9% /kidneys/ for the former and 6.2% /skeletal muscles/ and 12.7% /spleen/ for the latter.

The ASF of the organs under study also strongly differ in the total contents of nitrogenous bases and nucleosides. Thus the total amount of nitrogenous compounds in the ASF of the testes is 4.1 μM/100 g of the tissue and that of the lungs is 80.8 μM/100 g; nucleosides: 5.6 μM/100 g /brain/ and 65.5 μM/100 g /lungs/. The difference in the contents of mono-, di-, and triphosphates of nucleosides is less pronounced except for lung tissue where diphosphates and triphosphates of nucleosides are contained in much lower concentrations /33.3 and 23.7 µM/100 g of the tissue, respectively/ than in other organs.

Thus, when evaluating the above indices characterizing the pools of free purine and pyrimidine derivatives contained in the organs of intact rats, it is first of all noteworthy that each of these organs displays its individual composition of the ASF. In other words, the qualitative and quantitative parameters of the ASF of the tissues of normal organs are specific. It seems that the tissue specificity of normal, intact organs is of a regular nature and is defined by the functional activity and the level of the proliferative processes in the organ.

Time dependence of the composition of the acid-soluble fraction of the rat liver after the death of the animals

Table II presents the ASF composition data at different periods after the death of the animals.

Within 1 h after death, the total amount of the ASF components of the liver is not changed; however, the redistribution of the compounds can be noted. Thus, a decrease in the ATP content is accompanied by an equimolar growth of the ADP and AMP levels. A similar pattern is also observed with the guanine and hypoxanthine derivatives.

Within 2 h, the total amount of the nucleotide material grows considerably: from 610.8 to 1209.0 µM/100 g of the tissue. A process of dephosphorylation of nucleoside triphosphates can be noted. TMP is first identified in the ASF and the overall level of thymidine and cytidine derivatives increases almost by a factor of 4.

Within 6-8 h, the process of the decomposition of tri- and diphosphates of nucleosides becomes more pronounced. ATP, GTP, ADP, GDP, CDP, and CTP disappear from the ASF. At the same time the total amount of the ASF components reaches its maximum 239.7 µM/100 g by 6 h and the highest concentrations of uric acid are found.

Table II Time dependence of the composition of acid soluble fraction of rat liver after the death of the animals

Extraction and assay procedures were as described in "Materials and Methods". Data represent mean values of four experiments ± S.E.

Purine and pyrimidine contents / μmol/100 g wet wt/

	Norm	Time after death of the animals/hour/				
		1	2	3	4	6
Adenine	19.6±1.3	20.8±0.9	36.3±7.1	42.4±3.9	76.8±27.3	85.7±12.9
Adenosine	10.6±2.8	13.3±1.6	20.3±3.4	34.6±5.1	62.4±11.6	93.7±14.8
AMP	115.0±19.3	123.1±17.4	266.1±17.9	302.0±40.5	519.0±70.3	816.7±93.6
ADP	102.8±12.4	114.8±16.2	151.8±16.8	80.4±16.8	36.2±5.1	23.8±1.9
ATP	116.6±17.4	93.9±14.5	87.3±19.8	27.7±2.8	6.3±2.4	6.2±2.9
Total adenine derivat.	364.6±29.0	365.9±27.9	561.8±50.5	487.1±44.4	700.7±76.5	1026.1±95.7
Guanine	0	0	7.1±0.1	15.4±2.3	24.7±4.0	32.4±3.8
Guanosine	18.2±2.9	14.3±1.9	47.9±8.3	59.7±9.1	66.2±8.5	96.3±12.0
GMP	18.7±4.9	29.8±3.2	27.2±4.1	39.4±6.2	57.6±13.1	40.9±7.6
GDP	13.1±2,3	17.2±3.1	42.6±9.3	40.4±7.2	21.7±7.3	11.8±1.5
GTP	20.1±2.7	13.6±2.9	7.1±0.6	2.9±0.2	36.0±1.1	0
Total guanine derivat.	70.1±6.7	74.9±5.7	132.0±13.1	157.8±13.4	173.8±17.8	180.4±14.8
Hypoxanthine	9.9±0.7	8.5±1.2	34.8±2.6	71.9±11.5	97.3±10.8	106.3±16.6
Inosine	9.4±0.8	11.5±2.7	17.5±3.2	21.4±4.8	37.4±5.7	37.1±6.2
IMP	25.0±7.2	28.6±3.4	29.2±4.7	36.7±7.3	16.4±0.9	7.3±1.6
Total hypox. derivat.	58.8±7.3	59.4±4.5	91.3±6.3	137.1±14.5	153.2±12.3	150.7±17.8

Table II contd.

	Time after death of the animals /hour/				
	8	10	16	24	48
Adenine	120.2±19.1	144.8±16.3	176.5±21.8	199.1±19.6	92.3±14.7
Adenosine	71.9±13.3	116.2±26.5	123.9±18.6	91.5±8.4	56.2±7.3
AMP	432.0±29.6	120.3±17.2	183.0±20.7	26.9±3.8	7.5±1.2
ADP	17.3±2.6	5.4±0.8	0	0	0
ATP	0	0	0	0	0
Total adenine derivat.	641.4±37.7	386.7±35.6	483.4±35.3	317.5±21.7	156.0±16.5
Guanine	46.3±5.4	73.2±11.7	63.5±8.4	52.3±4.9	21.9±3.1
Guanosine	102.9±12.2	126.8±21.5	114.7±16.2	129.1±19.6	77.4±8.2
GMP	32.8±6.5	51.9±9.3	31.8±8.5	24.3±5.2	19.8±3.6
GDP	2.7±0.2	0	0	0	0
GTP	0	0	0	0	0
Total guanine derivat.	184.7±15.7	251.9±26.2	210.0±20.1	205.7±20.9	119.1±9.5
Hypoxanthine	118.9±19.4	123.0±16.5	138.5±21.3	136.2±15.8	191.9±35.3
Inosine	46.4±7.0	51.9±8.3	66.5±7.9	44.2±9.3	32.0±8.1
IMP	1.5±0.2	0	0	0	0
Total hypox. derivat.	166.8±20.6	174.9±18.5	205.0±22.7	180.4±18.3	223.9±36.2

Table II contd.

	Norm	Time after death of the animals /hour				
		1	2	3	4	6
XMP	4.4±0.6	6.1±0.3	16.8±3.1	21.5±4.7	36.5±9.2	62.8±3
Total PURINES	497.9±30.7	506.3±28.8	801.9±52.7	803.5±48.8	1064.2±78.1	1419.7±98.8
Uric acid	6.2±0.4	8.3±0.7	29.5±2.5	70.8±10.3	96.0±15.4	173.9±26.2
Thymine	0	0	0	6.3±2.1	8.2±1.9	32.5±8.5
Thymidine	7.5±1.3	6.8±0.9	20.8±7.3	60.3±14.6	237.0±48.1	256.1±40.1
TMP	0	0	3.6±0.9	9.1±1.7	16.0±2.8	30.9±5.8
Total thymine derivat.	7.5±1.3	6.8±0.9	24.4±7.4	75.7±14.9	261.2±48.2	319.5±41.4
Uracil	21.3±2.9	18.5±3.0	20.3±3.6	62.9±13.9	82.0±9.5	106.5±23.8
Uridine	0	0	36.6±8.2	41.5±9.6	48.1±8.4	53.8±12.5
UMP	22.5±3.1	24.8±2.1	92.3±25.1	104.1±32.6	20.6±4.2	9.2±2.8
UDP	21.4±6.3	20.5±4.7	135.3±39.3	15.4±2.6	8.6±2.5	0
UTP	15.7±3.8	17.2±2.3	10.5±1.1	4.2±0.7	0	0
Total uracil derivat.	80.9±8.5	81.0±6.4	294.0±47.5	223.9±60.1	159.3±13.6	169.5±27.0
Cytosine	0	0	8.3±1.4	60.2±11.9	79.0±10.0	110.3±16.2
Cytidine	0	0	12.1±2.6	140.3±15.3	158.2±22.1	140.5±18.6
CMP	7.5±1.5	5.9±0.7	30.5±5.8	88.7±9.5	95.9±16.3	60.3±6.1
CDP	10.8±2.9	12.1±1.6	8.3±1.2	4.9±0.7	0	0
Total cytosine derivat.	18.3±3.3	18.0±1.8	59.2±6.6	294.1±21.6	331.1±29.2	331.1±25.4
Total PYRIMIDINE	106.7±9.2	105.8±6.7	377.6±48.5	593.7±65.6	763.6±58.0	800.1±55.6
TOTAL COMPONENTS	610.8±32.1	620.4±29.6	1209.0±1.7	1468.0±82.4	1913.0±98.5	2393.7±116.4

Table II contd.

	Time after death of the animals/hour/				
	8	10	16	24	48
XMP	15.2 ± 1.4	3.5 ± 2.2	0	0	0
Total PURINES	1008.1 ± 45.8	817.0 ± 48.0	898.4 ± 55.3	695.6 ± 35.3	499.0 ± 40.9
Uric acid	101.3 ± 19.7	38.5 ± 4.8	36.7 ± 5.1	29.8 ± 3.6	30.5 ± 2.8
Thymine	44.1 ± 10.6	55.2 ± 13.1	68.2 ± 12.7	56.4 ± 9.3	55.8 ± 11.2
Thymidine	302.0 ± 38.4	217.0 ± 42.2	177.1 ± 14.5	99.3 ± 15.6	81.7 ± 9.8
TMP	25.1 ± 3.9	18.0 ± 2.5	7.7 ± 0.8	2.1 ± 0.3	1.6 ± 0.2
Total thymine derivat.	371.2 ± 40.0	290.2 ± 44.3	253.0 ± 19.3	157.8 ± 18.2	139.1 ± 14.9
Uracil	78.3 ± 6.6	32.4 ± 7.8	13.0 ± 4.2	10.0 ± 3.2	6.3 ± 0.9
Uridine	68.4 ± 11.6	53.2 ± 16.5	91.4 ± 13.8	44.8 ± 7.2	49.1 ± 5.3
UMP	0	0	0	0	0
UDP	0	0	0	0	0
UTP	0	0	0	0	0
Total uracil derivat.	146.7 ± 13.4	85.6 ± 18.3	104.4 ± 14.4	54.8 ± 7.9	55.4 ± 5.4
Cytosine	126.0 ± 12.7	153.7 ± 33.2	99.4 ± 13.8	66.9 ± 14.2	71.6 ± 9.5
Cytidine	98.2 ± 20.3	52.4 ± 14.4	36.0 ± 8.2	19.5 ± 3.8	20.0 ± 6.5
CMP	45.0 ± 12.5	29.3 ± 4.6	11.8 ± 2.3	0	0
CDP	0	0	0	0	0
Total cytosine derivat.	269.2 ± 27.0	235.4 ± 36.5	147.2 ± 16.2	86.4 ± 14.7	91.6 ± 11.5
Total PYRIMIDINE	787.1 ± 50.1	611.2 ± 60.3	504.6 ± 29.0	299.0 ± 24.7	286.1 ± 19.6
TOTAL COMPONENTS	1896.5 ± 70.7	1466.0 ± 772.3	1439.7 ± 62.7	1024.4 ± 43.2	815.6 ± 45.4

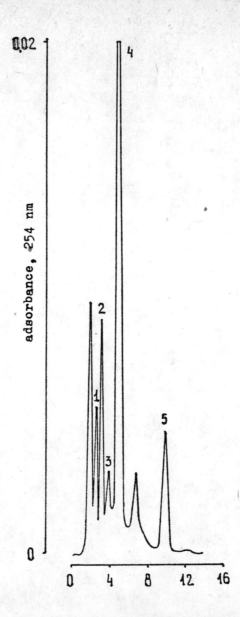

Fig. 4 Isocratic HPLC separation of the whole
zone of nucleoside monophosphates /zone B
in Fig. 1/ obtained on the Analyst 7800
instrument /Partisil-5 ODS/. 1, CMP; 2, UMP;
3, GMP, IMP; 4, AMP; 5, NAD$^+$.

Within 8 h, tri- and diphosphates decompose completely and the concentration of monophosphates decreases. By then the amount of pyrimidines starts falling.

Within 16 h, only nucleoside monophosphates, nucleosides, and nitrogenous bases can be determined in the composition of the ASF.

Within 48 h, only insignificant concentration of monophosphates, nucleosides, and nitrogenous bases are recorded together with a maximum level of hypoxanthine 191.1 µM/100 g . Cytidine and thymidine bases and nucleosides are retained on a relatively high level.

Thus, four stages of the ASF composition dynamics can be distinguished in the liver autolysis: within 1 h after death, the dephosphorylation of nucleoside triphosphates to di- and monophosphates is noted without any changes in other compounds; within 2-3 h, the dephosphorylation process is still more pronounced and the products of decomposition of nucleic acids start entering the ASF; within 6-10 h, a complete decomposition of di- and triphosphates of nucleosides and a maximum increase in the amounts of the RNA and DNA catabolism products in the ASF is observed; within 16 h, the rate of autolysis /endonucleolysis/ decreases and the number of the components drops.

The diurnal variation of the ASF composition of the mouse liver

The concentrations of purine and pyrimidine derivatives in the ASF of the mouse liver as a function of the time of the day are shown in Table III.

The total amounts of purine and pyrimidine derivatives vary differently: purines show a maximum at 20.00 and two minima, at 0.4.00 and 16.00 /$p < 0.05$/; pyrimidines manifest a sudden increase between 20.00 and 24.00 and a gradual reduction to minimum values between 04.00 and 16.00

Among pyrimidine derivatives, the levels of thymidine, TMP, and dCMP are of a definitely pulsed nature: maximum values are observed in the evening and at night and minimum values are observed in the morning and in the daytime. Uridine derivatives,

Table III Diurnal variation of purine and pyrimidine derivatives in the acid-soluble fraction of mouse liver

Extraction and assay procedure were as described in "Materials and Methods". Data represent mean values of five experiments ± S.E.

Purine and pyrimidine contents /μmol/100 g wet wt/

	10.00	12.00	14.00	16.00	18.00	20.00
Adenine	7.2±0.7	57.3±6.9	51.4±6.1	46.9±4.6	53.7±5.4	68.4±7.1
Adenosine	43.3±3.3	14.9±4.3	26.8±3.5	13.8±2.6	26.8±3.2	42.3±6.9
AMP	190.2±18.4	232.9±26.6	153.1±17.3	101.6±14.8	140.2±12.1	229.6±19.4
ADP	137.3±14.2	151.6±14.6	146.2±11.4	150.2±11.3	160.9±14.6	176.1±12.1
ATP	113.3±8.6	116.9±8.5	121.9±8.3	139.6±10.9	152.3±11.5	167.7±8.4
Total adenine derivat.	491.3±25.0	573.6±32.5	499.4±23.4	452.1±22.2	533.9±23.1	684.1±26.3
Guanine	4.5±0.2	0	0	0	0	0
Guanosine	12.2±0.7	10.6±1.8	9.7±2.1	11.8±2.6	7.3±1.9	8.8±2.2
GMP	26.9±3.4	28.8±7.4	25.3±6.3	22.6±6.9	19.7±5.4	17.8±4.4
GDP	0	0	0	0	0	0
GTP	25.8±2.7	37.9±8.7	31.8±7.9	23.4±7.5	25.1±6.2	28.2±4.1
Total guanine derivat.	69.4±4.4	77.3±11.6	66.8±10.3	57.9±10.5	52.1±11.4	54.8±9.4
Hypoxanthine	13.6±3.2	8.9±2.4	7.3±1.5	5.8±0.8	2.6±0.1	1.3±0.1
Inosine	37.7±9.8	30.7±4.1	30.2±3.6	31.6±2.3	23.5±2.8	19.2±2.6
IMP	36.2±3.9	37.8±5.6	47.5±4.9	41.0±6.1	36.2±3.1	33.7±3.8
Total hypox. deriv.	87.5±11.0	77.4±7.3	85.0±6.3	78.4±6.6	62.3±4.2	54.2±4.6

Table III contd.

	Purine and pyrimidine contents / μmol/100 g wet wt/					
	22.00	24.00	2.00	4.00	06.00	08.00
XMP	12.0±1.8	10.6±3.4	7.2±2.9	4.7±0.8	8.1±1.8	19.0±2.5
Total PURINES	660.2±27.7	738.9±35.4		593.1±25.5	656.4±26.2	812.1±28.4
Uric acid	4.7±0.9	0	0	0	0	6.7±1.8
Thymine	0	0	0	0	0	0
Thymidine	5.7±1.2	0	0	0	0	9.4±1.3
TMP	0	0	0	0	0	9.3±0.8
Total thymine derivat.	5.7±1.2	0	0	0	0	18.7±1.5
Uracil	2.6±0.1	3.2±0.2	4.6±0.3	3.9±0.2	7.5±0.3	8.2±0.4
Uridine	0	0	0	0	0	0
UMP	27.4±3.1	27.1±4.2	22.1±2.9	19.9±1.2	32.4±3.8	45.2±8.5
UDP	39.8±5.2	41.8±11.7	40.1±5.8	36.7±8.2	45.2±5.8	53.4±3.1
UTP	17.0±2.7	14.7±3.1	17.5±1.8	21.4±4.2	23.1±3.3	26.3±2.8
Total uracil derivat.	68.6±6.6	86.8±12.3	84.3±6.2	81.9±9.6	108.2±7.7	133.1±9.5
CMP + dCMP	7.9±1.1	0	0	0	0	5.1±0.4
CDP	3.3±0.2	3.7±0.2	0	1.6±0.8	4.2±0.2	6.5±0.3
Total cytosine deriv.	11.2±1.1	3.7±0.2	0	1.6±0.8	4.2±0.2	11.6±0.5
Total PYRIMIDINES	103.7±6.8	90.5±12.3	84.3±6.2	83.5±9.6	112.4±7.7	163.4±9.6
TOTAL COMPONENTS	768.6±28.5	829.4±37.5	742.7±27.2	676.6±27.2	768.8±27.3	975.5±30.0

Table III contd.

	Purine and pyrimidine contents / μmol/100 g wet wt/					
	22.00	24.00	02.00	04.00	06.00	08.00
Adenine	76.1±9.3	105.2±9.2	67.8±5.9	73.3±6.1	57.6±5.2	43.0±4.6
Adenosine	33.1±5.4	30.4±4.8	20.2±4.1	15.6±3.8	26.1±6.8	29.4±7.4
AMP	150.8±12.6	118.4±13.9	111.0±9.9	92.9±12.5	120.3±15.2	181.3±23.4
ADP	201.4±17.0	182.3±17.6	175.3±13.2	143.5±9.6	132.6±16.9	151.1±11.3
ATP	162.1±9.3	155.7±6.8	123.8±10.6	127.6±15.4	147.9±11.6	141.2±6.3
Total adenine derivat.	623.5±25.5	592.0±25.6	498.1±18.0	452.9±23.2	484.5±26.9	546.0±28.1
Guanine	0	0	0	2.8±0.1	3.6±0.1	4.2±0.1
Guanosine	12.7±2.5	14.3±3.1	11.8±2.5	9.3±2.7	10.4±2.9	11.3±2.2
GMP	15.1±3.5	12.9±4.1	9.1±2.3	6.0±1.1	18.0±3.7	21.0±5.7
GDP	0	10.6±5.1	8.0±1.4	6.2±1.8	0	0
GTP	30.8±5.6	29.8±7.2	32.3±6.7	30.0±4.3	25.4±3.8	24.0±5.7
Total guanine derivat.	58.6±9.2	67.6±16.3	61.2±12.5	53.3±9.1	57.4±8.9	60.5±11.6
Hypoxanthine	0	0	0	9.8±0.3	7.5±0.2	9.1±2.1
Inosine	17.1±2.2	20.7±3.4	23.7±3.6	21.3±1.8	2.0±4.2	41.0±3.9
IMP	20.3±2.7	26.6±2.4	20.8±5.1	53.0±5.7	60.5±7.7	47.2±7.1
Total hypox. derivat.	37.4±3.5	47.3±4.2	44.5±6.3	84.6±6.0	70.0±8.8	97.3±8.4

Table III contd.

Purine and pyrimidine contents / µmol/100 g wet wt/

	22.0	24.00	02.00	04.00	06.00	08.00
XMP	23.5±3.7	27.1±2.8	30.2±3.5	31.9±4.0	21.8±2.1	16.2±2.9
Total PURINES	743.0±27.6	734.0±30.9	634.0±23.0	622.7±26.0	633.7±29.8	720.0±31.7
Uric acid	15.4±2.1	21.3±3.2	34.8±5.7	29.5±5.3	10.9±1.1	0
Thymine	2.5±0.1	4.3±0.3	0	0	0	0
Thymidine	11.4±2.3	15.5±3.6	17.6±2.9	13.3±1.8	12.6±1.7	11.8±0.9
TMP	10.7±1.6	28.9±5.2	24.1±4.8	18.1±2.2	7.2±0.9	0
Total thymine derivat.	24.6±2.8	48.7±6.7	41.7±5.6	31.4±2.9	19.8±1.9	11.8±0.9
Uracil	92.1±1.2	15.3±0.9	14.7±2.3	10.5±1.6	7.0±0.4	3.1±0.2
Uridine	0	0	0	0	0	0
UMP	52.1±5.3	68.2±6.4	61.4±5.1	57.4±3.2	42.3±4.8	34.6±7.1
UDP	58.1±6.1	53.7±4.7	59.2±4.8	65.7±9.1	44.6±3.7	32.4±2.8
UTP	30.2±4.5	35.6±3.3	33.1±3.6	32.6±4.9	28.5±2.9	26.6±3.7
Total uracil derivat.	152.5±9.3	172.8±8.7	168.4±8.2	165.5±10.9	122.4±6.7	96.7±8.5
CMP + dCMP	8.1±0.9	24.6±1.4	16.2±2.3	28.9±2.3	12.4±1.6	0
CDP	6.9±0.5	9.3±0.5	13.7±1.9	17.8±0.9	8.3±0.9	5.6±0.3
Total cytosine derivat.	15.0±1.0	33.9±1.8	29.9±3.0	45.7±2.5	20.7±1.8	5.6±0.3
Total PYRIMIDINES	192.1±9.8	255.4±11.3	240.0±10.4	242.6±11.6	162.9±7.2	114.9±8.6
TOTAL COMPONENTS	935.1±29.3	989.4±32.9	874.0±25.3	865.3±28.5	796.6±30.7	834.9±32.8

constituting a considerable proportion of the whole ASF pool of the intact liver, also manifest considerable diurnal variations: their content starts growing at 20.00 and holds the highest level between 24.00 and 04.00.

Likewise, the concentrations of purine derivatives of the ASF of the intact liver vary markedly during a 24 hours' period. The drop of the concentrations of adenine and guanine derivatives at night may be due both to the activation of their decomposition into final products and to the inhibition of their biosynthesis. It is possible that the activation of the catabolism of purines predominated in the intact mouse liver, as confirmed by the accumulation of uric acid. Uric acid content is increased in the evening and at night /from 20.00 to 04.00/. During this period of time, the levels of IMP, AMP, and GMP are at minimum and the XMP content grows.

It is well known that the liver is the main organ performing the biosynthesis of purines for the organism, including purines for the haemopoietic bone marrow where the de novo synthesis of these compounds is virtually not effected /10/. The physiological functioning and reproduction of hepatocytes are mutually exclusive processes. Bearing this in mind, we may say that high levels of purine nucleotides detected in daytime favour the concept of diurnal nonuniformity of the organization of the biosynthesis of purine derivatives in the intact liver.

The relationships between purine and pyrimidine derivatives during their accumulation in different periods of the day seem to underly self-regulation of the processes of reparation /growth/ and physiologic functioning in the liver.

The composition of the ASF of
the regenerating mouse liver

The biosynthesis of DNA in the mouse liver regenerating after partial hepatectomy begins not earlier than 24-30 h after operation /11/. Within 4-6 h after partial hepatectomy, a sufficiently simultaneous transition of the first proliferating population of hepatocytes into the G_1 period occurs; thus,

from this particular instant and up to 24.00, the cells of
the liver parenchyma exist in the prereplicative phase and
after 24.00 in the replicative phase.

The most characteristic and distinctive features of the
nucleotide composition of the ASF of the regenerating liver in
the prereplicative stage is a strong decrease in the contents
of purine derivatives and a considerable rise of the levels of
pyrimidines /Table IV/.

The variation of the contents of all pyrimidine nucleotides within up to 20 h after hepatectomy is of a wave-life character. The highest amplitude of the variations was found for
thymidine and TMP. As early as 2-4 h after operation, the concentration of thymidine becomes 15 times higher than the initial level, and a maximum content of TMP is recorded. This rise
of the amounts of TMP and thymidine can be explained by haemo-
dynamic disturbances developing immediately after liver operation and followed by partial destruction of the hepatic cells
both in the suture zone and in the whole mass of the remaining
tissue.

Within 6-8 h after operation, events related to the induction of the enzymes of the biosynthesis of purines and pyrimidines, i.e. the preparation of replication as such, develop.
Intensified synthesis of uridine derivatives is noted.

The intensification of the RNA biosynthesis at the initial
stages of the development of proliferative processes is one of
the earliest changes detected in the regenerating liver /12,
13/. The dynamics of the concentrations of uridine nucleotides
observed in our studies is in a complete accordance with these
findings. In the course of the whole studied period, a persistent increase in the contents of these substances, particularly UTP, is detected. Maximum UTP levels were recorded within
2-4 and 16-20 h after partial hepatectomy, thus possibly pointing to the action of one of the regulatory mechanisms governing
the biosynthesis of pyrimidines. This is especially evident
within 10-20 h after operation when UDP appears in the ASF; before this instant, UDP is not detectable in the regenerating
liver. In its turn, high concentration of UMP /compared with
control/ reflects the activation of the synthesis of pyrimidines,

Table IV. Purine and pyrimidine derivatives in the

Extraction and assay procedures were as described in three experiments $\pm$ S.E.

	Sham oper-ated mice	Purine and pyrimidine		
		Time after		
		2	4	6
Adenine	8.6$\pm$0.7	7.3$\pm$0.4	4.8$\pm$0.2	5.3$\pm$0.4
Adenosine	48.8$\pm$4.3	82.2$\pm$6.7	65.4$\pm$4.8	43.0$\pm$3.8
AMP	211.3$\pm$28.4	182.4$\pm$12.1	164.5$\pm$10.8	189.8$\pm$14.5
ADP	148.8$\pm$17.3	102.6$\pm$8.5	75.8$\pm$10.1	96.8$\pm$7.4
ATP	136.6$\pm$9.9	46.6$\pm$4.3	31.7$\pm$3.8	46.5$\pm$5.6
Total adenine derivatives	554.1$\pm$35.0	421.1$\pm$16.8	342.1$\pm$16.0	373.4$\pm$17.6
Guanine	5.3$\pm$0.2	4.8$\pm$0.3	7.5$\pm$0.6	3.3$\pm$0.1
Guanosine	12.3$\pm$1.2	13.8$\pm$1.5	16.3$\pm$1.7	11.4$\pm$0.9
GMP	25.4$\pm$5.3	33.8$\pm$4.3	50.8$\pm$6.2	21.5$\pm$2.8
GDP	O	O	O	O
GTP	31.2$\pm$4.1	28.7$\pm$4.7	28.8$\pm$2.7	24.3$\pm$3.2
Total guanine derivatives	74.2$\pm$6.8	69.3$\pm$6.6	103.4$\pm$7.0	60.5$\pm$4.35
Hypoxanthine	12.7$\pm$3.5	O	O	O
Inosine	6.8$\pm$0.8	80.4$\pm$9.4	33.6$\pm$5.2	10.9$\pm$0.9
IMP	22.3$\pm$3.7	24.6$\pm$4.1	11.8$\pm$2.9	14.9$\pm$1.8
Total hypo-xanthine derivatives	41.8$\pm$5.2	105.0$\pm$10.2	45.4$\pm$6.0	25.8$\pm$2.0
XMP	12.6$\pm$0.9	26.3$\pm$1.8	57.2$\pm$5.3	11.4$\pm$2.1
Total PURINES	682.7$\pm$36.1	621.7$\pm$20.8	538.0$\pm$19.2	471.1$\pm$18.4
Uric acid	4.7$\pm$0.7	10.2$\pm$0.8	O	O

acid-soluble fractions of regenerating mouse liver

"Materials and Methods". Data represent mean values of

contents / µmol/100 g wet wt/				
hepatectomy /hour/				
8	10	16	20	24
9.3 ± 0.7	8.8 ± 0.6	12.6 ± 1.2	9.4 ± 0.5	7.3 ± 0.5
77.3 ± 4.5	85.6 ± 7.1	93.4 ± 4.6	66.6 ± 8.2	64.0 ± 5.2
170.3 ± 25.2	157.7 ± 13.4	148.4 ± 13.2	169.5 ± 17.8	111.5 ± 9.4
48.2 ± 11.2	42.7 ± 7.2	71.8 ± 4.3	62.9 ± 8.5	80.0 ± 7.6
22.5 ± 1.8	22.9 ± 3.9	43.5 ± 3.9	39.8 ± 5.0	46.0 ± 5.1
327.6 ± 28.0	317.7 ± 17.3	369.7 ± 15.2	348.2 ± 22.0	308.8 ± 14.1
9.3 ± 0.7	8.1 ± 0.5	6.8 ± 0.4	7.3 ± 0.3	2.4 ± 0.1
10.3 ± 0.8	13.3 ± 1.4	10.2 ± 0.9	8.7 ± 0.5	0
27.9 ± 2.9	22.4 ± 8.4	25.6 ± 4.2	28.4 ± 2.3	18.2 ± 2.3
0	0	0	0	0
19.0 ± 3.1	18.5 ± 1.9	25.0 ± 2.8	36.7 ± 3.4	20.8 ± 1.9
66.5 ± 4.4	62.3 ± 8.7	67.5 ± 5.2	81.1 ± 4.2	41.4 ± 3.0
$37.4\pm4,8$	$28.2\pm5,2$	0	0	0
13.7 ± 2.1	26.4 ± 2.7	96.0 ± 8.5	39.8 ± 5.6	21.6 ± 1.4
17.4 ± 1.4	20.8 ± 3.1	21.6 ± 2.9	44.5 ± 4.9	53.2 ± 6.7
68.5 ± 5.4	75.4 ± 6.6	117.6 ± 9.0	84.3 ± 7.5	74.8 ± 9.7
30.5 ± 3.6	26.7 ± 2.9	49.0 ± 4.6	12.2 ± 0.9	8.1 ± 0.3
493.1 ± 29.1	482.1 ± 20.7	603.8 ± 28.1	525.8 ± 23.6	433.1 ± 17.4
9.1 ± 0.7	9.1 ± 0.8	17.1 ± 2.1	0	0

Table IV

	Sham-oper- ated mice	Time after		
		2	4	6
Thymine	O	O	O	O
Thymidine	$6.2^{\pm}1.1$	$85.8^{\pm}9.4$	$68.2^{\pm}7.8$	O
TMP	O	$50.6^{\pm}4.2$	$25.5^{\pm}3.8$	O
Total thymine derivatives	$6.2^{\pm}1.1$	$136.4^{\pm}10.3$	$93.7^{\pm}8.7$	O
Uracil	$3.5^{\pm}0.1$	$2.8^{\pm}0.2$	$7.5^{\pm}1.1$	$3.6^{\pm}0.2$
Uridine	O	O	O	O
UMP	$30.6^{\pm}2.2$	$39.6^{\pm}5.1$	$44.4^{\pm}7.1$	$48.8^{\pm}4.5$
UDP	$36.4^{\pm}4.1$	O	O	O
UTP	$18.9^{\pm}2.6$	$19.4^{\pm}2.7$	$50.7^{\pm}4.7$	$61.2^{\pm}8.2$
Total uracil derivatives	$89.4^{\pm}5.3$	$61.8^{\pm}5.8$	$102.6^{\pm}8.5$	$113.6^{\pm}9.4$
Cytosine	O	O	$1.8^{\pm}0.1$	O
Cytidine	O	$3.4^{\pm}0.2$	$2.3^{\pm}0.2$	O
CMP	$4.4^{\pm}0.3$	$17.6^{\pm}1.2$	O	O
CDP	$3.5^{\pm}0.2$	$18.9^{\pm}1.6$	$8.9^{\pm}0.6$	$3.8^{\pm}0.2$
Total cytosine derivatives	$7.9^{\pm}0.4$	$39.9^{\pm}2.0$	$13.0^{\pm}0.7$	$3.8^{\pm}0.2$
Total PYRIMIDINES	$103.5^{\pm}5.3$	$238.1^{\pm}12.0$	$209.3^{\pm}12.2$	$117.4^{\pm}9.4$
TOTAL COMPONENTS	$790.9^{\pm}36.5$	$870.0^{\pm}24.0$	$757.3^{\pm}22.8$	$588.5^{\pm}20.7$

contd.

<table>
<tr><th colspan="5">hepatectomy /hour/</th></tr>
<tr><th>8</th><th>10</th><th>16</th><th>20</th><th>24</th></tr>
<tr><td>3.7 ± 0.2</td><td>6.9 ± 0.5</td><td>7.1 ± 0.3</td><td>11.6 ± 0.2</td><td>o</td></tr>
<tr><td>39.4 ± 4.1</td><td>75.4 ± 5.9</td><td>67.4 ± 8.5</td><td>43.4 ± 6.1</td><td>27.1 ± 0.9</td></tr>
<tr><td>26.8 ± 3.1</td><td>24.4 ± 2.2</td><td>27.9 ± 3.1</td><td>20.8 ± 3.5</td><td>13.2 ± 2.5</td></tr>
<tr><td colspan="5"></td></tr>
<tr><td>69.9 ± 5.2</td><td>106.7 ± 6.3</td><td>102.4 ± 9.1</td><td>65.8 ± 7.0</td><td>40.3 ± 2.7</td></tr>
<tr><td>4.2 ± 0.7</td><td>6.1 ± 0.3</td><td>3.7 ± 0.2</td><td>4.8 ± 0.3</td><td>5.3 ± 0.4</td></tr>
<tr><td>6.3 ± 0.4</td><td>12.7 ± 0.3</td><td>14.6 ± 1.2</td><td>16.0 ± 1.3</td><td>11.5 ± 0.8</td></tr>
<tr><td>59.8 ± 5.2</td><td>53.2 ± 3.5</td><td>58.9 ± 4.8</td><td>61.2 ± 7.8</td><td></td></tr>
<tr><td>o</td><td>o</td><td>o</td><td>39.8 ± 4.6</td><td>47.4 ± 3.8</td></tr>
<tr><td>35.3 ± 5.3</td><td>36.8 ± 3.9</td><td>57.7 ± 7.2</td><td>84.2 ± 6.7</td><td>73.2 ± 6.1</td></tr>
<tr><td colspan="5"></td></tr>
<tr><td>105.6 ± 7.5</td><td>108.8 ± 5.3</td><td>134.9 ± 8.7</td><td>206.0 ± 11.4</td><td>137.4 ± 7.3</td></tr>
<tr><td>o</td><td>o</td><td>2.6 ± 0.2</td><td>o</td><td>o</td></tr>
<tr><td>2.8 ± 0.3</td><td>4.6 ± 0.3</td><td>6.3 ± 0.5</td><td>o</td><td>o</td></tr>
<tr><td>16.9 ± 1.2</td><td>18.4 ± 0.9</td><td>33.5 ± 2.7</td><td>o</td><td>11.0 ± 1.7</td></tr>
<tr><td>10.6 ± 0.9</td><td>9.3 ± 0.8</td><td>17.3 ± 1.1</td><td>4.1 ± 0.7</td><td>8.3 ± 0.5</td></tr>
<tr><td colspan="5"></td></tr>
<tr><td>30.3 ± 1.5</td><td>32.3 ± 1.2</td><td>60.2 ± 3.0</td><td>4.1 ± 0.7</td><td>19.3 ± 1.8</td></tr>
<tr><td colspan="5"></td></tr>
<tr><td>205.8 ± 9.3</td><td>247.8 ± 8.3</td><td>227.5 ± 13.0</td><td>275.9 ± 13.4</td><td>197.0 ± 8.0</td></tr>
<tr><td colspan="5"></td></tr>
<tr><td>708.0 ± 30.6</td><td>739.0 ± 22.3</td><td>848.4 ± 31.0</td><td>801.7 ± 27.1</td><td>630.1 ± 19.1</td></tr>
</table>

Table V. Purine and pyrimidine derivatives in the acid-soluble.

Extraction and assay procedures were as described in three experiments ± S.E.

| Sham-operated mice | Purine and pyrimidine | | | |
| | Time after | | | |
	28	32	36	40
Adenine	10.6 ± 0.7	15.3 ± 0.9	3.5 ± 0.4	5.6 ± 0.7
Adenosine	51.2 ± 7.6	20.2 ± 2.7	10.5 ± 2.5	12.7 ± 1.8
AMP	120.2 ± 11.0	150.4 ± 12.3	169.8 ± 10.8	153.7 ± 14.2
ADP	96.3 ± 5.8	103.8 ± 7.6	120.3 ± 14.2	110.9 ± 8.4
ATP	72.3 ± 8.9	111.3 ± 9.9	129.5 ± 11.8	149.6 ± 15.5
Total adenine derivatives	350.6 ± 17.0	401.0 ± 17.8	433.6 ± 21.5	432.5 ± 22.7
Guanine	O	O	O	O
Guanosine	O	O	O	O
GMP	14.5 ± 0.7	19.4 ± 2.6	27.2 ± 3.5	33.6 ± 2.9
GDP	O	3.6 ± 0.2	5.9 ± 0.7	8.3 ± 0.9
GTP	9.6 ± 1.3	14.2 ± 2.3	21.6 ± 2.2	25.4 ± 3.7
Total guanine derivatives	24.1 ± 1.5	37.2 ± 3.5	54.7 ± 4.2	67.3 ± 4.8
Hypoxanthine	O	O	O	11.6 ± 0.9
Inosine	13.4 ± 1.9	19.8 ± 2.3	17.2 ± 2.8	25.4 ± 3.1
IMP	36.7 ± 8.5	41.5 ± 5.8	36.8 ± 4.5	43.4 ± 5.7
Total hypoxanthine derivatives	78.8 ± 9.5	93.6 ± 9.8	96.4 ± 7.5	110.8 ± 6.6
XMP	21.5 ± 3.1	26.2 ± 4.8	30.4 ± 3.9	22.3 ± 4.1
Total PURINES	475.0 ± 19.8	558.0 ± 21.1	615.1 ± 23.5	632.9 ± 24.5
Uric acid	O	O	O	O

fractions of regenerating mouse liver

"Materials and Methods". Data represent mean values of

contents (mol/100 g wet wt)				
hepatectomy (hour)				
44	54	72	96	168
12.3 ± 1.6	9.6 ± 0.8	3.5 ± 0.2	17.2 ± 2.6	12.4 ± 1.3
29.4 ± 2.5	32.3 ± 4.1	40.5 ± 7.2	39.5 ± 2.9	44.5 ± 5.6
120.8 ± 13.5	106.8 ± 11.2	87.2 ± 9.3	97.2 ± 7.4	98.7 ± 14.1
136.0 ± 19.4	150.1 ± 16.3	147.5 ± 12.8	163.5 ± 15.7	173.8 ± 19.4
159.5 ± 14.7	178.9 ± 21.3	200.7 ± 19.2	226.8 ± 19.6	212.6 ± 20.2
458.0 ± 28.0	477.6 ± 40.6	479.4 ± 40.6	479.4 ± 25.9	533.0 ± 31.9
0	0	1.4 ± 0.2	3.6 ± 0.3	4.8 ± 1.1
0	2.6 ± 0.3	3.1 ± 0.2	7.4 ± 0.5	11.5 ± 0.9
20.9 ± 1.8	23.6 ± 4.9	26.7 ± 3.2	20.1 ± 2.6	24.8 ± 3.5
2.1 ± 0.3	0	0	0	0
29.6 ± 4.3	30.2 ± 2.9	22.4 ± 3.8	30.1 ± 4.2	34.8 ± 5.2
52.6 ± 4.7	56.4 ± 5.7	53.4 ± 5.0	61.2 ± 5.0	75.9 ± 6.4
17.3 ± 2.5	14.8 ± 2.6	17.0 ± 1.3	15.4 ± 4.6	10.3 ± 2.5
41.7 ± 5.1	30.2 ± 6.7	23.5 ± 3.2	12.0 ± 2.1	5.5 ± 0.4
33.9 ± 4.1	26.5 ± 5.3	20.4 ± 1.9	27.3 ± 2.8	$24,7 \pm 3.5$
127.6 ± 7.0	94.3 ± 8.9	$78.4 \pm 4,0$	73.8 ± 5.8	55.8 ± 4.3
11.4 ± 0.9	7.7 ± 0.5	6.2 ± 1.2	5.0 ± 0.7	5.2 ± 0.3
649.6 ± 29.3	636.1 ± 42.0	617.2 ± 26.7	684.2 ± 27.6	669.9 ± 32.8
0	2.2 ± 0.3	3.6 ± 0.4	4.8 ± 0.3	5.6 ± 0.7

Table V

Sham-operated mice	Time after			
	28	32	36	40
Thymine	0	0	0	0
Thymidine	$30.2^{\pm}3.5$	$39.7^{\pm}4.1$	$44.5^{\pm}6.7$	$56.3^{\pm}4.9$
TMP	$10.6^{\pm}1.6$	$21.5^{\pm}3.2$	$30.7^{\pm}5.1$	$39.3^{\pm}4.2$
TTP	0	$2.3^{\pm}0.3$	$6.9^{\pm}0.5$	$7.8^{\pm}0.4$
Total thymine derivatives	$40.8^{\pm}3.9$	$64.1^{\pm}5.2$	$84.8^{\pm}8.4$	$109.2^{\pm}6.5$
Uracil	$2.6^{\pm}0.5$	0	0	0
Uridine	$14.8^{\pm}0.6$	$3.7^{\pm}0.3$	$2.1^{\pm}0.3$	$1.8^{\pm}0.1$
UMP	$23.0^{\pm}3.2$	$45.2^{\pm}9.9$	$41.6^{\pm}4.7$	$32.8^{\pm}5.8$
UDP	$54.3^{\pm}7.3$	$66.5^{\pm}11.2$	$50.1^{\pm}6.3$	$45.5^{\pm}5.9$
UTP	$99.4^{\pm}12.0$	$110.9^{\pm}13.5$	$80.2^{\pm}8.6$	$52.1^{\pm}7.2$
Total uracil derivatives	$184.1^{\pm}14.4$	$226.3^{\pm}20.2$	$174.0^{\pm}11.7$	$131.2^{\pm}11.0$
Cytosine	0	0	0	0
Cytidine	0	0	0	0
CMP	$12.6^{\pm}1.5$	$22.7^{\pm}2.9$	$19.3^{\pm}1.2$	$29.3^{\pm}3.8$
CDP	$10.5^{\pm}1.7$	$13.3^{\pm}2.8$	$11.8^{\pm}2.1$	$7.6^{\pm}0.9$
CTP	$2.3^{\pm}0.3$	$7.7^{\pm}0.5$	$1.4^{\pm}0.2$	0
Total cytosine derivatives	$25.4^{\pm}2.3$	$43.7^{\pm}4.1$	$32.5^{\pm}2.4$	$36.9^{\pm}3.9$
Total PYRIMIDINES	$250.3^{\pm}15.1$	$334.1^{\pm}21.3$	$291.3^{\pm}14.6$	$277.3^{\pm}13.4$
Total COMPONENTS	$725.3^{\pm}24.9$	$892.1^{\pm}30.0$	$906.4^{\pm}27.7$	$910.2^{\pm}39.3$

contd.

hepatectomy (hour)				
44	54	72	96	168
O	O	O	O	O
$66.0^{\pm}5.9$	$32.7^{\pm}4.1$	$12.3^{\pm}1.5$	$5.6^{\pm}0.3$	$2.3^{\pm}0.1$
$47.8^{\pm}3.3$	$28.3^{\pm}1.9$	$0.8^{\pm}0.2$	O	O
O	O	O	O	O
$124.3^{\pm}6.8$	$66.1^{\pm}4.5$	$13.1^{\pm}1.5$	$5.6^{\pm}0.3$	$2.3^{\pm}0.1$
O	$0.9^{\pm}0.1$	$1.5^{\pm}0.3$	$2.8^{\pm}0.3$	$4.7^{\pm}0.5$
O	O	O	O	O
$30.1^{\pm}2.5$	$33.3^{\pm}3.9$	$39.8^{\pm}2.6$	$36.1^{\pm}4.2$	$31.5^{\pm}4.8$
$39.1^{\pm}6.7$	$44.6^{\pm}8.4$	$37-8^{\pm}5.5$	$30.3^{\pm}4.1$	$26.3^{\pm}2.8$
$44.2^{\pm}5.3$	$36.8^{\pm}2.9$	$21.1^{\pm}3.3$	$13.8^{\pm}0.5$	$9.6^{\pm}0.7$
$113.4^{\pm}8.9$	$115.6^{\pm}9.7$	$100.2^{\pm}6.9$	$83.0^{\pm}5.9$	$72.1^{\pm}5.6$
O	O	O	O	O
O	O	O	O	O
$27.6^{\pm}4.1$	$7.8^{\pm}1.8$	$3.2^{\pm}0.5$	$3.9^{\pm}0.8$	$2.2^{\pm}0.3$
$5.4^{\pm}0.2$	O	O	O	O
O	O	O	O	O
$33.0^{\pm}4.2$	$7.8^{\pm}1.8$	$3.2^{\pm}0.5$	$3.9^{\pm}0.8$	$2.2^{\pm}0.3$
$270.7^{\pm}12.0$	$189.5^{\pm}10.9$	$116.5^{\pm}7.1$	$92.5^{\pm}6.0$	$76.6^{\pm}5.6$
$920.3^{\pm}31.7$	$827.8^{\pm}43.4$	$737.3^{\pm}27.6$	$781.5^{\pm}28.2$	$752.1^{\pm}33.3$

as confirmed by the growth of the total amount of uridine derivatives within 8 h after operation and by the build-up of the levels of CMP and CDP.

The dynamics of purine derivatives shows another pattern. First of all, within 2-20 h after operation, a deficiency of IMP and adenine nucleotides, most clearly expressed for ATP, arises. The content of ATP steadily decreases to 1/3-1/6 of the normal level. Concurrently with this, the concentrations of final and intermediate products of metabolism of purines vary quite dissimilarly. Within the first 2 h after operation, inosine accumulates in large quantities: its level increases more than 3 times. At the same time the contents of adenosine and XMP increase substantially. Another rise of the inosine level occurs only by 16 h. Within 8-16 h after operation, both the accumulation of XMP and of the products of catabolism of purines /uric acid, hypoxanthine/ and a persistent decrease of the concentration of IMP, the precursor of the AMP and GMP synthesis, are observed.

Sixteen hours after operation is an instant which seems to mark the beginning of the S period since from this particular instant until 20-24 h after operation, a decrease in the concentrations of thymidine and cytidine nucleotides is recorded, being caused by the beginning and intensification of the DNA synthesis. Within 24-28 h, an increase in the amounts of TMP, CMP, CDP, and thymidine in the ASF is again noticed which is possibly due to the activation of anabolism as a result of the consumption of final products.

It is noteworthy that TTP and CTP appear in the ASF within 32-36 h; within 40-44 h, when the maximum concentration of triphosphate is recorded, a decrease in the levels of all nucleotides starts which continues to 168 h when the normal level is reached /Table V/.

To sum up, the character of the variation of the concentrations of purine and pyrimidine derivatives in the ASF of the regenerating liver corresponds to several phases of the development of the process: ischemic, destructive changes with the activation of the pathways of the catabolism of purines and pyrimidines /from 2 to 6 h/; activation of the biosynthesis of nucleotides /from 6 to 16-20 h/; pronounced biosynthesis of the precursors of nucleic acids with the utilization of triphosphates of nucleosides in the DNA biosynthesis corresponding to the S period of the process /from 16-20 to 28-32 h/.

Between 28-32 and 44 h /period when an expressed mitotic activity in the regenerating liver is noted/, the ASF displays maximum levels of all the precursors of nucleic acids.

Thus, we can state the existence of stereotypes of pools of purine and pyrimidine derivatives corresponding to the processes of destruction and hypertrophic regeneration in the liver of the animals.

Comparative characterization of the ASF composition

To make comparative evaluations and to find out correlative links between different components of a complex assembly of substances found in each experimental point, we have introduced basic characteristics of the ASF composition: the total amount of substances, the overall content of nitrogenous bases, nucleosides, and nucleotides /mono-, di-, and triphosphates/, the percentages of the derivatives of each nitrogenous base, and the K_1-K_7 values. The K_1-K_7 coefficients have been defined as follows:

$$K_1 = \frac{\Sigma \text{ Purines}}{\Sigma \text{ Pyrimidines}} \; ; \; K_2 = \frac{AMP + 2ATP}{2/AMP + ADP + ATP/} \; ;$$

$$K_3 = \frac{\Sigma \text{ Adenine derivatives}}{\Sigma \text{ /Hypoxanthine derivatives} + XMP + \text{Uric acid/}} \; ;$$

$$K_4 = \frac{\Sigma \text{ Pyrimidines}}{\text{Uric acid}} \; ;$$

$$K_5 = \frac{\Sigma \text{ Bases} + \Sigma \text{ Nucleosides} + \Sigma \text{ Nucleoside monophosphates}}{\text{Nucleoside diphosphates} + \text{Nucleoside triphosphates}} \; ;$$

$$K_6 = \frac{\Sigma \text{ Adenine derivatives}}{\Sigma \text{ Pyrimidines}} \; ;$$

$$K_7 = \frac{\Sigma \text{ Adenine derivatives}}{\Sigma \text{ Hypoxanthine derivatives}}$$

These characteristics have been determined in the course of analysis of our experimental results except for K_2, the energy charge, according to Atkinson /14/ and make it possible to evaluate, on the whole, the direction of the metabolism of purines and pyrimidines and to detect individual distinctions in each separate case.

Thus, a comparison of four coefficients /K_1, K_2, K_3, K_4/ characterizing the ASF of the organs of intact rats shows that the characteristics of the ASF compositions of the organs with high polyfunctional activity /kidneys and liver/ and of the organs with expressed proliferative activity /testes and spleen/ are similar although specific differences also exist /Table VI/.

A comparative evaluation of the ASF coefficients of the intact and regenerating liver as well as of the liver after death confirms the idea about the existence of stereotype pools of purine and pyrimidine derivatives peculiar to definite morphofunctional states of the tissue /Table VII/.

Method of comparative chromatograms

A complete combined analysis of the ASF composition is very time-consuming and requires relatively large amounts of the material: organ tissues or blood. At present, the methods of high-performance liquid chromatography featuring good reproducibility and sensitivity make the analysis much simpler and faster since they do not require maximum possible separation of the ASD components or determination of a complete set of quantitative characteristics. Under strictly fixed conditions, one may obtain chromatograms with a sufficiently informative /for clinical purposes/ profile, in other words, comparative chromatograms /2, 15/. For instance, the comparative chromatograms

Table VI Characteristics of ASF Composition of the
Rat Organs

Organs	K_1	K_2	K_3	K_4
Liver	4.6	0.5	5.3	17.2
Kidneys	5.4	0.59	2.4	5.6
Spleen	2.3	0.7	5.1	-
Muscular tissue	5.6	0.6	3.1	3.4
Lungs	4.1	0.25	1.4	5.6
Brain	5.6	0.8	10.7	-
Testes	4.6	0.2	46.9	-

Table VII Characteristics of the Pools of Free Purine
and Pyrimidine Derivatives at Different
States of the Liver Tissue

Stereo-type	Low level of functional and proliferative activity	High level of functional activity	High level of proliferative activity	Destruction	
Coefficients	norm	time /h/		regeneration /32-48 h after operation/	6-hours storage after sacrificing
	10.00	14.00 - 16.00	2.00- 4.00		
K_1	6.4	7.8	2.3	2.4	1.3
K_2	0.5	0.5	0.5	0.5	0.04
K_3	5.3	5.0	3.3	3.2	1.5
K_4	17.2	-	12.8	12.5	4.6
K_5	1.0	0.8	0.9	1.0	14.7
K_6	4.7	5.9	1.7	1.9	0.9
K_7	4.2	5.8	4.9	4.4	2.7

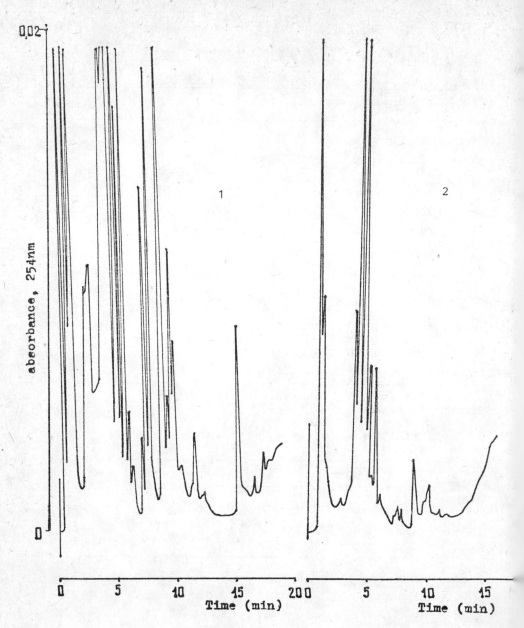

Fig. 5   Comparative chromatograms of the acid-soluble
fraction of liver /1/ and lungs /2/. Column:
Partisil-10 SAX /4.6 x 25 mm/. Mobile phase:
A - 0.01 M $NH_4H_2PO_4$, pH 3.8; B - 0.5 M
$NH_4H_2PO_4$; 0.5 M KCl, pH 4.3; 0 - 100%, concave curve, 20 min. Flow rate: 1.5 ml/min.
Injection volume: 200 µl.

of the intact organs of rats demonstrate the organ specificity of the ASF /Figs 5, 6, 7/.

We have used the method of comparative chromatograms in the analysis of the ASF of erythrocyte, plasma, and the tissues of the mucous membrane and the tumour in patients with carcinoma of the stomach. To make comparative chromatograms, we obtained about 5-10 mg of the tissues of the mucous membrane and the tumour of the stomach during gastroscopic examination and also samples of erythrocytes and plasma from venous blood.

A visual comparison of the chromatograms of a morphologically unchanged mucous membrane, the mucous membrane adjacent to the tumour /with histological signs of malignancy of the cells/, and of the tumour tissue as such /Fig. 8/ reveals quite clearly the differences in the chromatographic profiles.

The comparative chromatograms of erythrocytes /Fig. 9/ and blood plasma /Fig. 10/ are even more demonstrative. A visual analysis of the comparative chromatograms of erythrocytes in patients with gastritis and carcinoma of the stomach and also of normal plasma and plasma of individuals with carcinoma of the stomach is a convincing example of the importance of this method in differential diagnostics. We believe that the use of comparative chromatograms is most valuable in early diagnostics of the cancerous degeneration of the mucous membranes of the stomach and the duodenum since the chromatographic profile of the mucous membrane adjacent to the tumour is practically identical with that of the cancerous tissue. Even when no definite tumour growth is disclosed and only the beginning of the process is suspected, the analysis of biopsy samples of the mucous membrane performed by comparative chromatography may detect the malignancy signs. The comparative chromatograms may be of unquestionable significance in differential diagnostics of the cancerous degeneration of gastric and duodenal ulcers.

The comparative chromatograms of the ASF of erythrocytes and blood plasma may become a basis of a method of early recognition of tumour formation in general, regardless of the location and type of the tumour.

Thus, purine and pyrimidine derivatives are an object of interesting and promising studies with relation both to the

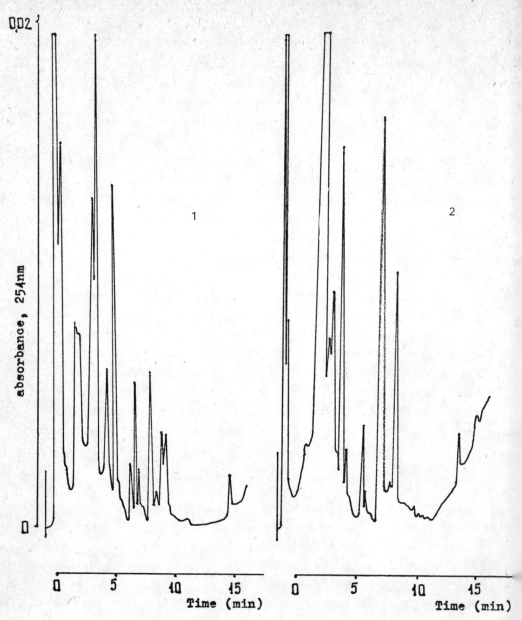

Fig. 6 Comparative chromatograms of the acid-soluble
fraction of kidneys /1/ and spleen /2/. For conditions see Fig. 5.

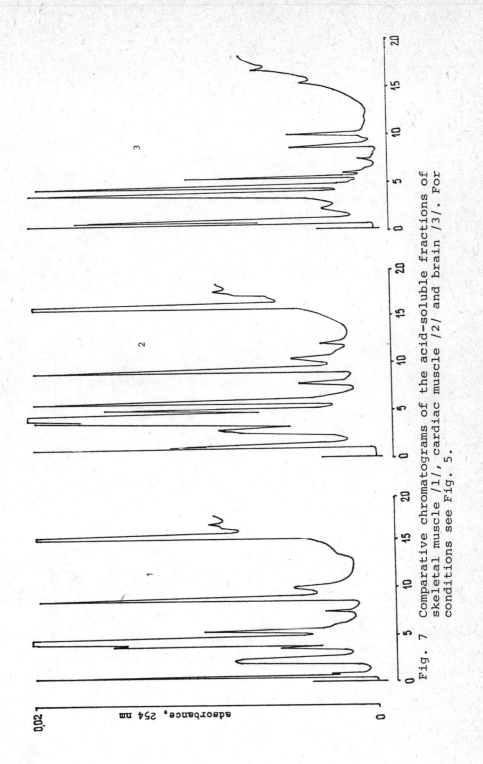

Fig. 7 Comparative chromatograms of the acid-soluble fractions of skeletal muscle /1/, cardiac muscle /2/ and brain /3/. For conditions see Fig. 5.

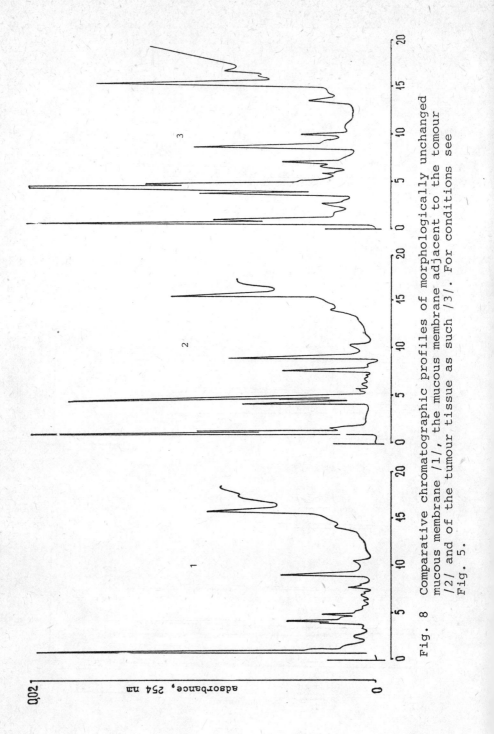

Fig. 8 Comparative chromatographic profiles of morphologically unchanged mucous membrane /1/, the mucous membrane adjacent to the tomour /2/ and of the tumour tissue as such /3/. For conditions see Fig. 5.

adsorbance, 254 nm

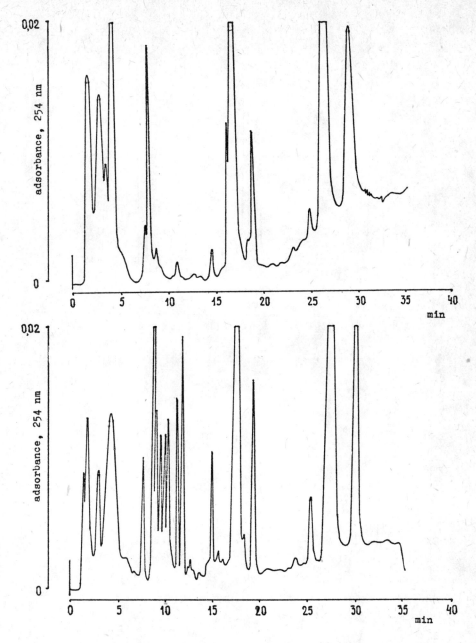

Fig. 9 Comparative chromatographic profiles of
        erythrocytes of patients with gastritis
        /upper panel/ and carcinoma of the stomach
        /lower panel/. For conditions see Fig. 5,
        except the time of gradient elution /40 min/.

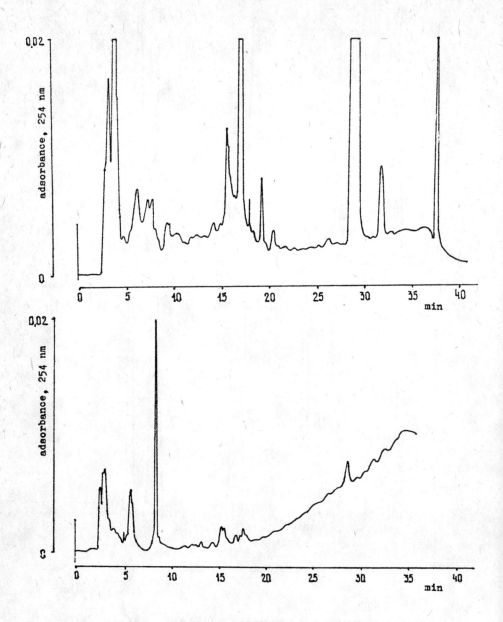

Fig. 10 Comparative chromatographic profiles of
blood plasma of patients with carcinoma
of the stomach /upper/ and normal subject
/lower/. For conditions see Fig. 5, except
the time of gradient elution /40 min/.

theory and practice. A clear understanding of the mechanism and control of the metabolism of purine and pyrimidine derivatives will broaden our knowledge of one of the basic aspects of the vital activity of the cell; on the other hand, qualitative and quantitative determinations of the compositions of the pools of purine and pyrimidine derivatives may be a basis of the development of simple and efficient methods of diagnostics of pathologic states.

REFERENCES

1. Y. Konishi and A. Ichihara, J. Biochem., 85 /1979/, 295.
2. P.R. Brown, R.A. Hartwick and A.M. Krstulovic, Biological /Biomedical Applications of Liquid Chromatography, Marcel Dekker, New York, 1979, p. 295.
3. T.L. Riss, N.L. Zorich, M.D. Williams and A. Richardson, J. Liq. Chromatogr., 3 /1980/, 133.
4. Y.M. Rustum, Anal. Biochem., 90 /1978/, 289.
5. E. Harmsen, J.W. de Jong and P.W. Serruys, Clinica Chimica Acta, 115 /1981/, 73.
6. R.C. Jackson, M.S. Lui, J. Boritzki, H.P. Morris and G. Weber, Cancer Researc h, 40 /1980/, 1286.
7. M.B. Cohen, J. Maybaum and W. Sadee, J. Chromatogr., 198 /1980/, 435.
8. G.M. Higgins and R.M. Anderson, Arch. Pathol., 12 /1931/, 186.
9. N.E. Hoffman and I.C. Liao, Anal. Chem., 49 /1977/, 2231.
10. A.M. Mackinon and D.S. Deller, Biochim. Biophys. Acta, 319 /1973/, 1.
11. I.N. Chernozemski and G.P. Warwick, Cancer Research, 30 /1970/, 2685.
12. E. Bresnick, J. Biol. Chem., 240 /1965/, 2550.
13. N.L.R. Bucher and M.N. Swaffield, Biochem. Biophys. Acta, 186 /1969/.
14. D.E. Atkinson, Biochem., 7 /1968/, 4030.
15. A.M. Krstulovic, R.A. Hartwick and P.R. Brown, Clinica Chimica Acta, 97 /1979/, 159.

ION EXCHANGER OSTION KS 4.2% AND SEPARATION OF SACCHARIDES

J. COPIKOVÁ, F. KVASNICKA and V. MUSIL*

Department of Saccharide Chemistry and Technology, Institute
of Chemical Technology, 166 28 Praha 6, 3, Suchbátarova
*Spolek pro chemickou a hutni vyrobu, Usti n. Labem,
 Czechoslovakia

INTRODUCTION

 With the growing appreciation of investigation of starch
syrup, more and more pressure is being placed on introducing
a rapid method for the separation and quantitation of different
mono- and oligosaccharides in a single analysis.

 Efficient separations of saccharides have been achieved by
partition chromatography on bonded-phase silica using aqueous
mixtures of acetonitrile or ethanol as the eluent. However, the
solubility of oligomers in aqueous acetonitrils or ethanol mixture decreases as the degree of polymerization /DP/ increases.
Gel chromatography on Bio-Gel P-2 with water as the eluent gives
effective separations of oligosaccharides up to DP of 18 but
is time-consuming and does not yield sufficient separations of
monosaccharides. Partition chromatography on cation exchange
resin in calcium form has been successfully used for the separation of mono- and oligosaccharide mixtures. Using the cation-
exchange resin OSTION LG KS 0402 /4.2% cross-linked resin/ we
were successful in separating oligosaccharides derived from potato starch up to a degree of polymerization of 8. We have used
this cation-exchange resin for the analysis of potato starch
syrup as well as for the determination of the content of glucitol /sorbitol/, D-mannitol, saccharose, fructose and glucose
in diabetic foodstuffs. Different monosaccharides and oligosaccharides up to DP of 8 were resolved within 35 min in   a
single run with water as the mobile phase. The column is suitable for routine sugar analysis over prolonged periods.

EXPERIMENTAL

Materials

The chromatographic resin OSTION LGKS 0402 /$\overline{dp}$ = 11 µm/, 4% cross-linked, was obtained from Spolek Pro Chemickou a Hutni Vyrobu /Ústi n.L. ČSSR/. The calcium chloride used for the preparation of the resin was reagent grade, from Lachema n.p. /Brno, ČSSR/.

Chromatographic-grade tubing /30 cm x 0.78 I.D./ and chromatographic end-fittings were obtained from Mikrotechna n.p. /Praha, ČSSR/.

Monosaccharides including saccharose, maltose and malto-triose were reagent grade, from Koch-Light Laboratories Ltd. /Colnbrook, England/. Acid hydrolyzates of starch were commercial preparations.

Instrumentation

The liquid chromatographic system used in this study consisted of a differential refractometer RIDK 101, a membrane pump MP 2501, a low pressure part LPP 01, a line recorder TZ 4100 /all Laboratorni pristroje, n.p., Praha, ČSSR/ and computing integrator SP 4100 /Spectra-Physics, Santa Clara, U.S.A./. Samples were introduced into the column with a microsampling valve /Mikrotechna n.p., Praha, ČSSR/.

Sample preparations

Since salts, acid, soluble protein and particular matter interfere with the quantitative analysis of samples they were removed prior to analysis. Samples were therefore cleaned up with solutions of Carrez I /$ZnSO_4$/ and Carrez II /$K_4Fe/CN/_6$/ and treated with mix-bed resin Amberlite MB-3.

Chromatographic conditions

Solvent, deionized, degassed water maintained at 95°C; glass column dimensions, 30 /resp. 20/ cm x 0.78 cm I.D.; column temperature, 80°C; detector attenuation, x 32; recorder sensitivity, 10 mV full scale; sample concentration, 0.1 - 2% dry solid basis; sample volume, 60 μl.

RESULTS AND DISCUSSION

Fig. 1 shows the separation of an acid-hydrolyzed potato starch syrup, 78.8% - dry solid basis; 39 DE /Dextrose Equivalent/ on a commercially prepared 4.2% cross-linked resin /Ca^{2+}/, run at the flow rate of 16.5 ml.h^{-1}. The column was 30 cm x 0.78 cm I.D. Table I illustrates the reduction in the number of theoretical plates for one meter column length due to changes in the flow rate.

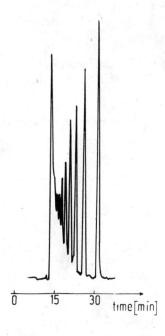

Fig. 1   Chromatogram of a 39 DE acid-hydrolyzed potato starch syrup. Chromatographic conditions: column; 30 cm x 0.78 cm I.D.; Ostion LG KS /Ca^{2+}/, 4.2% cross-linked resin /$\overline{dp}$ = 11 μm/; flow rate 16.5 ml.h^{-1}, pressure 1 MPa.

811

We verified by the calculation of the response factor RF /RF = $\frac{\text{peak area}}{\text{concentration}}$; for glucose 3.2×10^6, for maltose 3.0×10^6, for maltotriose 3.1×10^6/ that the glucose and maltodextrins have the same area response on a mass basis. Therefore the column may be calibrated with either glucose or maltose. The average oligosaccharide content up to DP of 8 of an acid-hydrolyzed potato starch syrup is listed in Table II.

Table I Number of theoretical Plates of the Column /30 cm x 0.78 cm I.D./ for Maltose

Flow rate ml.h^{-1}	Pressure MPa	Number of theoretical plates per meter column length $/\text{m}^{-1}/$
16	0.9	11 500
16.5	1	12 000
17	1.15	11 500
17.5	1.3	11 000
18	1.45	10 800

Table II The Average Content of Saccharides in an Acid-hydrolyzed Potato Starch Syrup /78.8%, 39 DE/ and Retention Times /Column 30 cm x 0.78 cm I.D., flow rate 16.5 ml.h^{-1}.

Syrup derived saccharides	Retention time /t_R/ /min/	Concentration /wt-%/
oligo-	13.9	24.3
DP - 8	16.5	3.3
DP - 7	17.4	5.0
DP - 6	18.3	6.4
DP - 5	19.8	7.6
DP - 4	21.5	9.5
DP - 3	23.6	11.8
DP - 2	26.6	14.0
DP - 1	31.9	18.1

We attempted to use the same resin for the determination
of saccharides in diabetic food-stuffs. The chromatogram   of
a carbohydrate mixture is shown in Fig. 2.

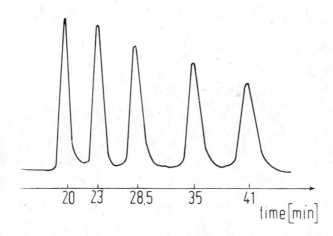

Fig. 2   Chromatogram of a carbohydrate mixture. Chromatographic
conditions: Column, 20 cm x 0.78 cm I.D.; Ostion LG KS
/Ca^{2+}/, 4.2% cross-linked resin /$\overline{dp}$ = 11 μm/; flow
rate 15 ml.h^{-1}: pressure 0.8 MPa.

t_R = 20 min - saccharose; t_R = 23 min - glucose,
t_R = 28.5 min - fructose; t_R = 35 min - D-mannitol;
t_R = 41 min - glucitol

We were interested in the determination of alditols as
glucitol /sorbitol/ and D-mannitol. We evaluated the calibration equations from the calibration curves:

for glucitol $Y = 10.3 \cdot 10^6 X - 0.05$ /1/,

for D-mannitol $Y = 11.3 \cdot 10^6 X - 0.02$ /2/,

Y .... peak area /integrator counts/

X .... concentration of alditol /%/.

With help of the equations /1 resp. 2/ we were able to
establish the content of glucitol and D-mannitol in diabetic
cherry and apple canned fruit /Fig. 3/, diabetic hard caramels
and diabetic lemon juice /Fig. 4/ and in diabetic milk chocolate and diabetic bitter chocolate. Our results are given in
Table III.

The separated carbohydrates were identified on the basis of their retention times and co-chromatography by thin-layer chromatography. Each analysis took only 45 minutes.

Table III. Saccharide content of Diabetic Foodstuffs

	saccharose	glucose	fructose	glucitol	mannitol
apple-canned fruit	0.1	1.2	3.6	0.1	-
cherry-canned fruit	0.2	3.2	2.7	0.7	-
lemon juice	0.04	0.6	1.1	2.0	0.1
hard caramels	-	-	-	93.0	4.2
bitter chocolate	0.7^x	3.7	-	36.3	0.9
milk chocolate	9.3	-	0.1	32.0	0.6

^{x}lactose

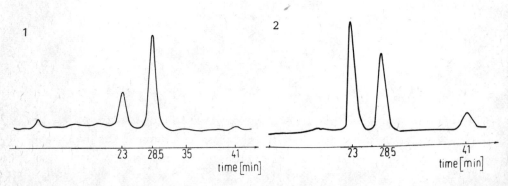

Fig. 3 Chromatograms of diabetic canned fruit samples.
Chromatographic conditions : see Fig. 2

1... apples;

t_R= 23 min /glucose/,

t_R= 28.5 min /fructose/,

t_R= 35 min /D-mannitol/,

t_R= 41 min /glucitol/

2... cherry

1

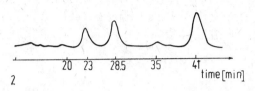

2

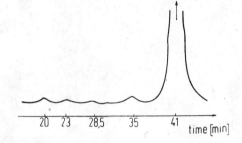

Fig. 4 Chromatograms of
diabetic-lemon juice
and diabetic hard
caramels. Chromato-
graphic conditions: see
Fig. 2

1... caramels;

t_R= 23 min
/glucose/,
t_R= 28.5 min
/fructose/,
t_R= 35 min
/D-mannitol/,
t_R= 41 min
/glucitol/

2... lemon juice;
t_R= 20 min
/saccharose/,
t_R= 23 min
/glucose/
t_R= 28.5 min
/fructose/
t_R= 35 min
/D-mannitol/
t_R= 41 min
/glucitol/

CATION EXCHANGE SEPARATION OF YTTRIUM IN MIXED SOLVENTS

A.G. GAIKWAD and S.M. KHOPKAR

Department of Chemistry, Indian Institute of Technology,
Bombay-400 076, India

Although several methods are known for the cation-exchange
chromatographic separations of yttrium (1), no work has been
carried out so far for the cation-exchange separation of
yttrium from mixed solvents.

Thorium was separated from yttrium (2) with dimethyl sulph-
oxide and formic acid as eluents. Europium was separated from
yttrium (3) with various non-aqueous solvents. The systematic
investigations on the distribution coefficient of yttrium in
mixed solvents containing hydrochloric (4) or nitric acid led
to the development of some methods for the separation but such
methods were never directed towards separation from commonly
associated elements. Therefore such studies on cation-exchange
behaviour of yttrium in mixed solvents are reported in this
paper.

EXPERIMENTAL

Apparatus and Reagents:

Ion-exchange column (1.4x20 cm), Dowex 50 W-X8 (Dow Chemical Co., Midland, Mich., USA) 20-50 mesh (H$^+$ form).

Digital pH meter type PM-822 (ECIL, India) with glass and
calomel electrodes.

Stock solution of yttrium: it was prepared by dissolving
1.60 g of yttrium oxide (Indian Rare Earths Ltd., Alwaye) in
250 ml of distilled water containing 5% nitric acid. It was
standardized complexometrically (5): as found it contained
5 mg/ml of yttrium.

Table I. Elution behaviour of yttrium

Y = 12.50 mg   Column 1.4 x 20 cm

Eluant (M)	Peak elution volume	Total elution volume	Yttrium recovery	Elution constant	Volume distribution coefficient	Weight distribution coefficient
	V_{max}, ml	V_t, ml	(%)	(E)	(D_v)	(D_w)
HCl						
3	100	275	100.0	0.39	2.55	1.77
4	75	200	100.0	0.58	1.74	1.21
5	50	200	98.2	1.08	0.93	0.65
HNO_3						
4	100	250	100.0	0.39	2.55	1.77
5	75	225	100.6	0.58	1.74	1.21
6	75	200	94.3	0.58	1.74	1.21
H_2SO_4						
3	50	250	100.0	1.08	0.93	0.65
4	50	200	100.0	1.08	0.93	0.65
$HClO_4$						
3	–	250	–	–	–	–
4	–	250	45.0	–	–	–
NH_4Cl						
3	125	300	95.0	–	–	–
4	75	150	100.0	0.58	1.74	1.21
5	50	125	100.0	1.08	0.93	0.65
NaCl						
3	100	300	96.0	0.39	2.55	1.77
4	75	250	100.0	0.58	1.74	1.21
5	50	225	100.0	1.08	0.93	0.65

towards various eluting
agents

Eluant (M)	Peak elution volume V_{max}, ml	Total elution volume V_t, ml	Yttrium recovery (%)	Elution constant (E)	Volume distribution coefficient (D_v)	Weight distribution coefficient (D_w)
$NaNO_3$						
5	125	350	100.0	–	–	–
6	100	300	100.0	0.39	2.55	1.77
$(NH_4)_2SO_4$						
0.5	75	200	95.0	0.58	1.74	1.21
1	50	100	100.0	1.08	0.93	0.65
2	50	100	100.5	1.08	0.93	0.65
CH_3COONH_4						
0.5	75	125	100.0	0.58	1.74	1.21
1	50	100	100.0	1.08	0.93	0.65
Malonic acid pH 4.4 5%	50	100	100.0	1.08	0.93	0.65
Tartaric acid pH 6 5%	50	150	100.0	1.08	0.93	0.65
EDTA pH 6 0.01M	50	100	100.0	1.08	0.93	0.65

GENERAL PROCEDURE

An aliquot of solution containing 12.5 mg of yttrium was
sorbed on the column at a flow rate of 1 ml/min. The column was
washed with water. From the column yttrium was eluted with
either various mineral acids (Table I) or a mixture of 4M hydro-

chloric acid with various concentrations of non-aqueous solvents
as the eluants (Table II). The effluent lot was collected in
25-ml fractions. Each fraction was evaporated almost to dryness,
and from each fraction yttrium was determined by complexometry
(5).

RESULTS AND DISCUSSION

The elution constant (E) and volume and weight distribution coefficients (D_V, D_W) were evaluated (6). The selectivity
scale for eluants was

H_2SO_4 > CH_3COONH_4 > $(NH_4)_2$ SO_4 >   Malonic acid >

Tartaric acid > EDTA > NH_4Cl > HCl   NaCl > HNO_3 > $NaNO_3$

However, for the mixed solvents, the selectivity scale was

Dioxane > 2-Propanol > Tetrahydrofuran > Ethanol >
Methanol >   Acetone.

Among the other eluants tested, malonic, tartaric acids
or EDTA were also effective. Among the acids sulphuric acid
was found to be the best but perchloric acid was a poor eluant.
Hydrochloric acid had practical utility from the point of rapid
determination. Among salts ammonium acetate and ammonium sulphate were also effective. Sodium nitrate, ammonium chloride and
sodium chloride are useful in higher concentration (Table I).
The behaviour of yttrium in the presence of 20-60% of various
organic solvents showed that dioxane and 2-propanol and tetrahydrofuran were efficient eluants in low concentration (20%)
but methanol, ethanol and acetone were poor eluants in higher
concentration (60%) (Table II, Fig 1).

Table II. Cation exchange studies of yttrium in mixed solvent systems

Yttrium = 12.50 mg. Weight of resin = 12.55 g. Dowex 5OW-X8 (H^+ form)

Organic solvents in % with 4M HCl	Peak elution volume $V_{max,ml}$	Total elution volume $V_{t,ml}$	Yttrium recovery (%)	Elution constant (E)	Volume distribution coefficient (D_v)	Weight distribution coefficient (D_w)
4M HCl	75	200	100.0	0.58	1.74	1.21
Methanol						
20	75	250	99.8	0.58	1.74	1.21
40	125	300	80.5	-	-	-
60	-	300	0.0	-	-	-
Ethanol						
20	75	250	99.5	0.58	1.74	1.21
40	100	300	83.2	0.39	2.55	1.77
60	-	300	0.0	-	-	-
2-Propanol						
20	75	250	99.9	0.58	1.74	1.21
40	75	300	88.6	0.58	1.74	1.21
60	-	300	0.0	-	-	-
Acetone						
20	75	250	99.5	0.58	1.74	1.21
40	75	300	52.5	0.58	1.74	1.21
60	-	300	0.0	-	-	-
1,4 Dioxane						
20	75	250	99.9	0.58	1.74	1.21
40	75	300	80.2	0.58	1.74	1.21
60	-	300	0.0	-	-	-
THF						
20	75	250	98.2	0.58	1.74	1.21
40	75	300	65.2	0.58	1.74	1.21
60	-	300	0.0	-	-	-

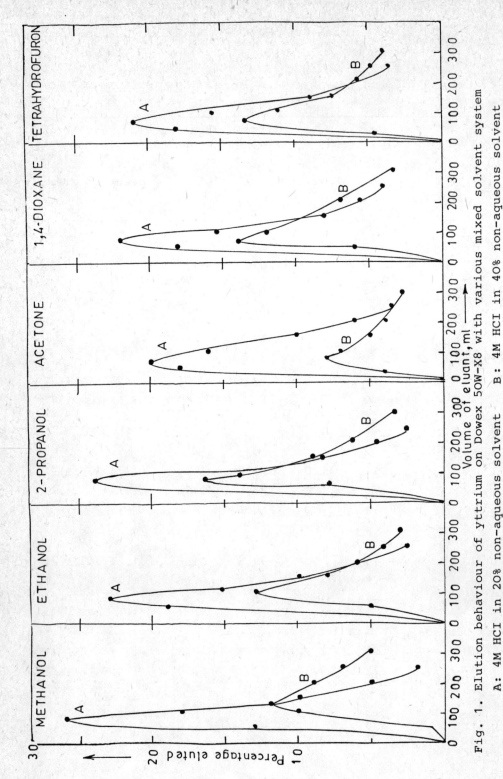

Fig. 1. Elution behaviour of yttrium on Dowex 50W-X8 with various mixed solvent system

A: 4M HCl in 20% non-aqueous solvent   B: 4M HCl in 40% non-aqueous solvent

SEPARATION OF YTTRIUM FROM BINARY MIXTURES

As yttrium was strongly bound to the resin in relation to other elements it was possible to first elute the loosely bound element with dilute hydrochloric acid containing appropriate proportion of non-aqueous solvents. Yttrium was eluted later with 4M hydrochloric acid.

In binary mixture manganese, cobalt, nickel and uranium(VI) were eluted first with 3 M hydrochloric acid in 60% tetrahydrofuran and yttrium was eluted later with 4M hydrochloric acid.

Similarly in binary mixtures zinc, indium, lead and bismuth were separated from yttrium by eluting these ions with 0.25M hydrochloric acid in 80% 1,4-dioxane, followed by elution of yttrium with 4M hydrochloric acid.

Aluminum, calcium, thorium and yttrium were separated by eluting aluminum with 3M hydrochloric acid in 60% methanol, calcium with 1.5M nitric acid, then thorium with 1M sulphuric acid, finally yttrium with 4M hydrochloric acid.

Uranium(VI), zirconium, thorium and yttrium were separated by eluting uranium(VI) with 1M hydrochloric acid in 80% acetone, zirconium with 0.5M sulphuric acid, thorium with 1M sulphuric acid and finally yttrium with 4M hydrochloric acid (Table III, Figs 2 to 6).

ANALYSIS OF YTTRIUM FROM MONAZITE

About one gram of monazite sand was digested by repeated treatment with concentrated sulphuric acid. It was diluted to 100 ml with 0.1M hydrochloric acid. An aliquot of the solution was taken and it was passed on the column at a flow rate of 1 ml/min. After sorption, aluminum was eluted with 3M hydrochloric acid in 60% methanol, calcium with 1.5M nitric acid, thorium with 1M sulphuric acid, cerium with 5% citric acid, and finally yttrium was eluted with 4M hydrochloric acid. Yttrium was found to be 1.75% against the standard value of 1.80%.

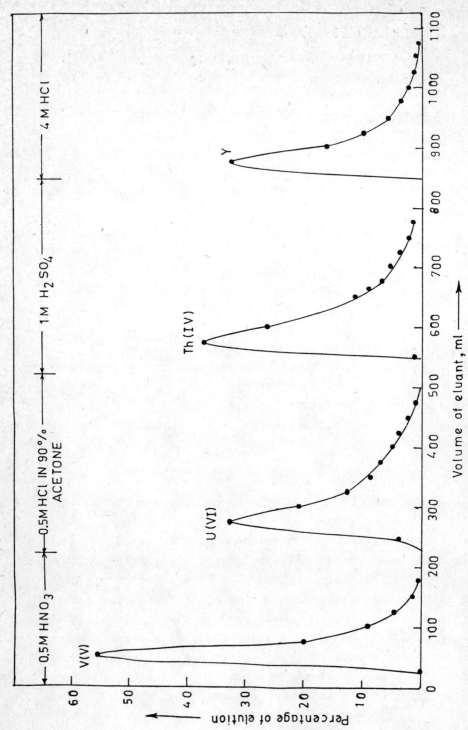

Fig. 2. Sequential separation of vanadium (V), uranium (VI) thorium (IV) and yttrium

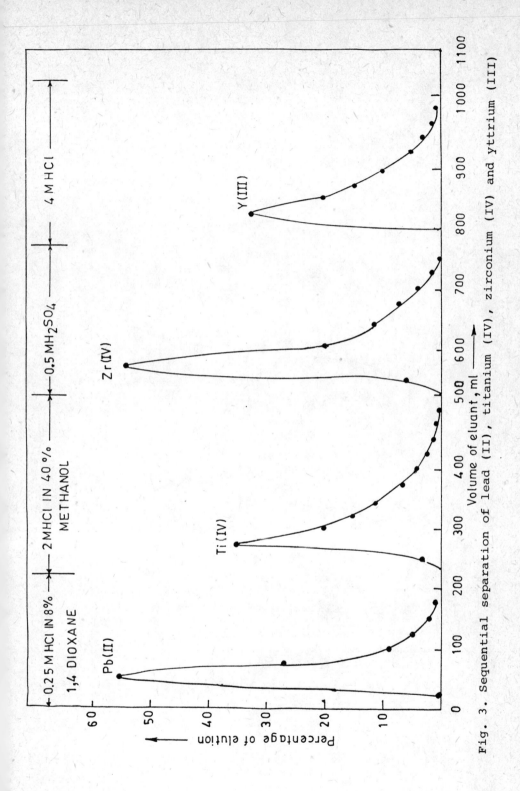

Fig. 3. Sequential separation of lead (II), titanium (IV), zirconium (IV) and yttrium (III)

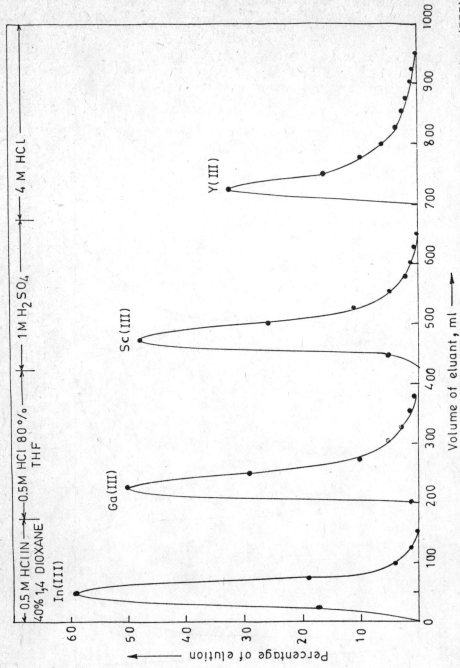

Fig. 4. Sequential separation of indium (III), gallium (III) scandium (III) and yttrium (III)

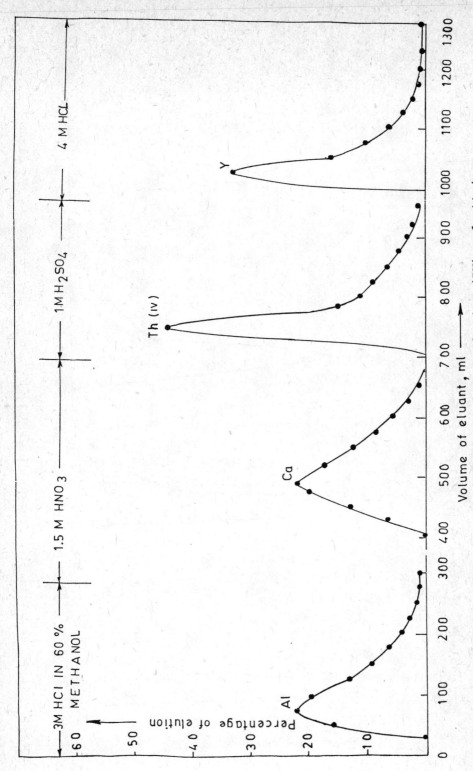

Fig. 5. Sequential separation of aluminum, calcium, thorium (IV) and yttrium

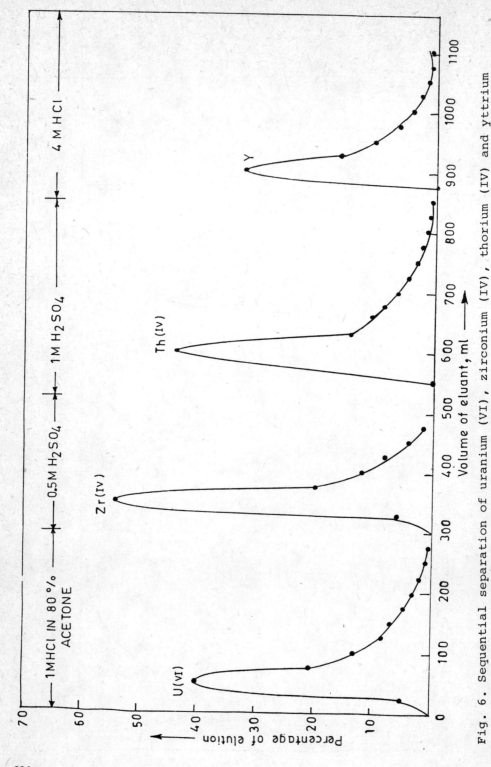

Fig. 6. Sequential separation of uranium (VI), zirconium (IV), thorium (IV) and yttrium

Table III. Cation exchange separation of yttrium from multi-component mixtures

No.	Elements	Amount added mg	Amount found mg	Recovery (%)
1.	V(V)	12.50	12.50	100.0
	U(VI)	13.50	13.50	100.0
	Th	10.50	10.51	100.1
	Y	8.50	8.48	99.8
2.	Pb	12.50	12.50	100.0
	Ti	10.50	10.49	99.9
	Zr	9.50	9.51	100.1
3.	In	12.0	12.0	100.0
	Ga	10.0	10.0	100.0
	Sc	12.50	12.51	100.1
	Y	8.50	8.49	99.9
4.	Al	10.50	10.50	100.0
	Ca	10.00	9.98	99.8
	Th	10.50	10.50	100.0
	Y	8.50	8.48	99.8
5.	U(VI)	13.50	13.50	100.0
	Zr	9.50	9.50	100.0
	Th	10.50	10.49	99.9
	Y	8.50	8.50	100.0

The total volume of eluant used throughout was 250 ml. All elements were determined by complexometric titrations using suitable indicator. The separation of yttrium from zirconium, uranium, thorium and titanium is important as they are usually associated in fission products. The separation of yttrium from scandium, aluminum, calcium and titanium is of significance as they can be found in minerals like monazite. The time required

Table IV. Cation exchange separation of yttrium from other elements

Yttrium = 8.50 mg

Foreign ions	Amount added mg	Amount found yttrium mg	Recovery of yttrium
Mn	20.50	8.50	100.0
Co	30.50	8.50	100.0
Ni	25.00	8.45	99.4
U(VI)	20.00	8.50	100.0
Zn	40.00	8.50	100.0
In	45.00	8.50	100.0
Pb	50.50	8.50	100.0
Bi	45.00	8.50	100.0
Hg(II)	50.00	8.50	100.0
Sn(IV)	45.00	8.49	99.9
As	48.00	8.48	99.8
Sb	51.00	8.50	100.0
Te(IV)	45.00	8.50	100.0
Fe(III)	25.00	8.49	99.9
Cu	26.00	8.48	99.8
Cd	25.50	8.50	100.0
Ga	20.00	8.50	100.0
Be	25.00	8.47	99.6
Ti(IV)	22.00	8.50	100.0
Mg	30.00	8.48	99.8
Al	20.00	8.50	100.0

for separation and determination is three hours. The method is simple, rapid and reproducible.

The separation of mercury(II), tin(IV), arsenic, antimony and tellurium(IV) from yttrium in a binary mixture was achieved by the elution of all these ions with 0.5M hydrochloric acid in 80% acetone and subsequent elution of yttrium with 4M hydrochloric acid.

After sorption of the binary mixtures, iron(III), copper, cadmium and gallium were eluted with 2M hydrochloric acid in 80% ethanol or beryllium and titanium(IV) were eluted with 2M hydrochloric acid in 40% methanol, or magnesium and aluminum were eluted with 3M hydrochloric acid in 60% methanol, and finally yttrium was eluted with 4M hydrochloric acid (Table IV).

SEPARATION OF YTTRIUM FROM A MULTICOMPONENT MIXTURE

The separation of vanadium(V), uranium(VI), thorium and yttrium was effected after passing the mixture on the column, by eluting vanadium(V) with 0.5M nitric acid, uranium(VI) with 0.5M hydrochloric acid in 90% acetone, thorium with 1M sulphuric acid, and yttrium with 4M hydrochloric acid.

The separation of lead, titanium, zirconium and yttrium was accomplished after sorption by eluting lead with 0.25M hydrochloric acid in 80% dioxane, titanium with 2M hydrochloric acid in 40% methanol, zirconium with 0.5M sulphuric acid and yttrium with 4M hydrochloric acid.

The trivalent ions such as indium, gallium, scandium and yttrium were separated after sorption by eluting indium with 0.5M hydrochloric acid in 40% dioxane, gallium with 0.5M hydrochloric acid in 90% tetrahydrofuran, scandium with 1M sulphuric acid. Finally yttrium was eluted as usual with 4M hydrochloric acid.

SUMMARY

Cation-exchange behaviour of yttrium was studied on Dowex 50W-X8 with various proportions of non-aqueous solvents in conjunction with 4M hydrochloric acid. Thus 1,4-dioxane, 2-propanol and tetrahydrofuran proved to be efficient eluants while methanol, ethanol and acetone were useful only at low concentrations. The separation of yttrium from a binary mixture was possible by exploiting the difference in the affinity of the metals for the resin. As yttrium was strongly bound, other ions were eluted first and yttrium was eluted later with 4M hydrochloric acid. The separation of yttrium from a multicomponent mixture was also possible. The method was extended for the analysis of yttrium from monazite sand.

REFERENCES

1. V.P. Mehta and S.M. Khopkar, Chromatographia, 11, 536(1978).

2. M. Qureshi and H. Khadim, Anal. Chim. Acta, 57, 387(1971).

3. J. Alexa, Colln. Czech. Chem. Commun, 30, 2344(1965).

4. F.W.E. Strelow, C.R. Vanzyl and C.J.C. Bothma, Anal. Chim. Acta, 45, 81(1969).

5. F.J. Welcher, Analytical application of ethylenediamine tetraacetic acid. D. Van Nostrand and Co., London (1958), P.175.

6. A.G. Gaikwad and S.M. Khopkar, J. Ind. Chem. Soc. 60, 1276(1982).

CHROMATOGRAPHICAL DETERMINATION OF ABSCISIC ACID AND ITS GLUCOSYL ESTER

H. LEHMANN

Institute of Biochemistry of Plants, Academy of Sciences of the GDR, Halle (Saale), Weinberg, GDR

INTRODUCTION

The plant hormone abscisic acid /ABA/ is involved in many important processes during plant growth and development such as induction of seed and flower dormancy, inhibition of elongation, promotion of abscission, closure of stomata and stress defence. ABA occurs in the plant not only in free form but also in conjugated forms such as glucosyl ester or glucosides. Among these the β-D-glucopyranosyl ester /ABA-Glc/ represents the main ABA conjugate. This compound is widely distributed in the plant kingdom and seems to be a permanent attendant of ABA. The physiological role of ABA-Glc is not yet clear: it may be a storage form of ABA from which the acid can be liberated by enzymatical processes or it may be the final product of ABA metabolism without any possibility of reconversion.

For investigations of the action of ABA and ABA-Glc in different processes during the plant development the exact determination of both compounds is necessary.

Quantitative analysis of ABA in plant material have been performed in general by physical methods /ORD, GLC/FID, GLC/ECD, GLC/MS, HPLC/UV/. These analytical procedures require a high degree of sample purification since low amounts of ABA are accompanied by large amounts of other organic and inorganic substances. There are some procedures of ABA purification mentioned in the literature, however, in general, the plant material is homogenised in 80% methanol. By extraction with alkaline methanol /1/ ABA-Glc may be converted to ABA methyl ester. The alkaline catalyzed transesterification starts immediately after addition of a NaHCO$_3$ solution to ABA-Glc dissolved in methanol /Fig. 1/ and yields large amounts of ABA methyl ester. Since ABA-Glc is also instable in aqueous alkaline solution /hydrolysis to free ABA/ the treatment of ABA-Glc with alkaline material has to be avoided.

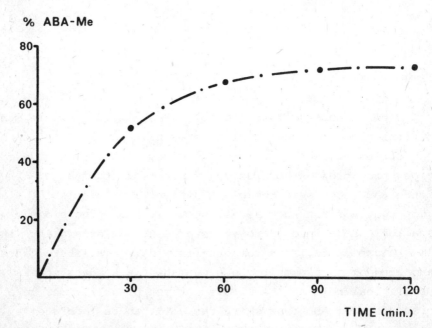

Fig. 1   Transesterification of ABA-Glc.
Addition of 0.1 ml saturated NaHCO$_3$ solution to
1 ml methanolic solution of ABA-Glc.

The separation of free and conjugated ABA during the different purification procedures has been carried out by partitioning of the water phase obtained after removing of methanol. This mode of separation may be also a source of errors. If ethyl acetate is used /2-5/ both free ABA and ABA-Glc are extracted. After four times of partition not only the total amount ABA is removed from the aqueous phase but also about 50% of ABA-Glc /Fig. 2/. Similar to ethyl acetate, extraction with n-butanol /6/ is not useful for the separation of ABA

and ABA-Glc, because the partition coefficient for ABA-Glc
/1.4/ is much higher than that observed in ethyl acetate /0.2/.
The remaining aqueous phase after partition against ethyl acetate or n-butanol was then used for the detection of conjugated
ABA dry hydrolysis. Since the ABA-Glc was partly extracted with
the organic solvents the results obtained cannot reflect true
values.

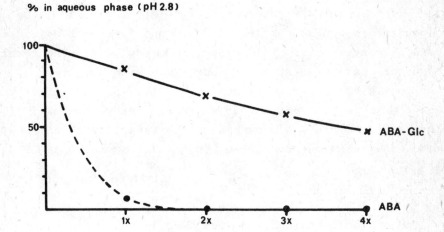

Fig. 2   Extraction of ABA and ABA-Glc by ethyl acetate.
          Equal volumes of water solutions and ethyl acetate
          were used.

Before ABA analysis by GLC many purification procedures involve TLC for ABA separation /1-11/. After TLC the ABA band is
scraped off and eluted. The loss of ABA during this step can
be high, up to 80%, and varies from experiment to experiment
performed under the same conditions. Consequently, if ABA is
purified by TLC for quantitative analysis an internal standard
for recovery calculation has to be an absolute necessity.
    The aim of this work was to improve the purification pro-
cedure for ABA analysis in order to prevent losses of ABA and

ABA-Glc as well as to avoid alteration of ABA-Glc during the different steps.

MATERIAL AND METHODS

Extraction procedures

10 g lyophilized plant material /or 100 g fresh material/ was homogenized in 100 ml /or 300 ml/ 80% methanol. After filtration and washing of the residue the methanolic solution was evaporated to aqueous phase. Pigments and fats were partly removed by centrifugation /20 min, 15,000 rev./min/. The remaining water phase was extracted four times with a mixture of ethyl acetate : n-butanol : n-propanol /50 : 10 : 5/ and, after acidification at pH 2.8, again four times with ethyl acetate. The ethyl acetate solution was washed with water, added to the ethyl acetate/n-butanol/n-propanol extract and evaporated at 35°C.

Column chromatography

The residue obtained was dissolved in 0.75 ml methanol : n-hexane : water /100 : 20 :3/ for chromatography on the first column.

1. Column:

DEAE-Sephadex A-25 /acetate form/ /2.8 x 19 cm/

Eluent: a. Methanol: n-hexane:water /100:20:3/ 150 ml
b. A mixture of 92.6 ml a + 7.4 ml acetic acid

Fraction size: 200 drops /4.8 ml/
ABA-Glc fractions: 22-28
ABA fractions: 52-56

The fractions containing ABA were collected, evaporated and dissolved in 1,2-dichloroethane : methanol /2:1/ for the second chromatography.

2. Column:

        Sephadex LH-20 /1.8 x 40 cm/
     Eluent: 1,2-dichloroethane : methanol /2:1/
     Fraction size: 200 drops /4.8 ml/
     ABA fractions: 14-17

Hydrolysis of ABA-Glc

The hydrolysis of ABA-Glc was performed with 1 N NaOH for
1 h at 80%. After acidification at pH 2.5 the free ABA was
extracted four times with ethyl acetate. The organic solution
was washed with water, evaporated and purified as described
above.

Calibration of the columns

Both columns were calibrated using radiolabelled standards:
/2-^{14}C/ABA : 63.6 MGq/m mol /synthesis according to ref. 12/.
/2-^{14}C/ABA-Glc : 63.6 MBq/m mol /synthesis according to ref.
13/. Because the ion exchanger cannot be regenerated in the
column, one separate column, working under the same conditions,
is necessary to determine the fractions in which ABA-Glc or
ABA occur.

RESULTS AND DISCUSSION

The method described /Fig. 3/ avoids purification steps
which cause substantial losses of ABA or ABA-Glc as well as
alteration of the substances. For separation of ABA and ABA-
-Glc and for the purification of ABA, naturally occurring or
obtained after hycrolysis, only column chromatography was used.
The plant material was homogenized in aqueous methanol. After
filtration and evaporation the remaining water phase was centrifuged and extracted with a mixture of ethyl acetate/n-buta-
nol/n-propanol. Both ABA and ABA-Glc are extractable by this
mixture in high yield /Fig. 4/.

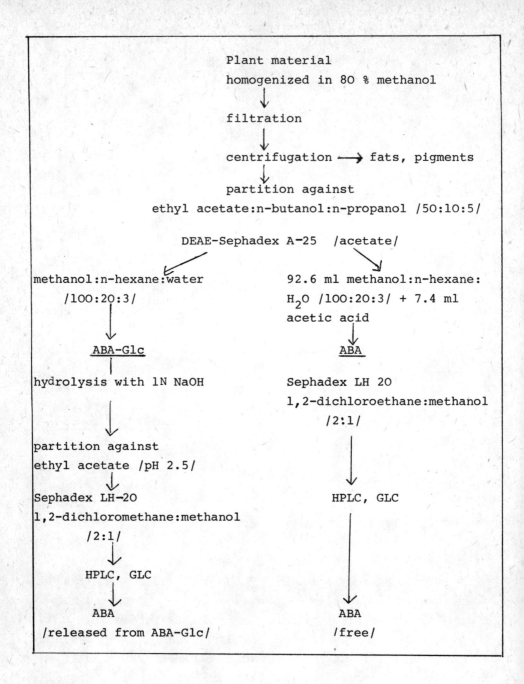

Plant material
homogenized in 80 % methanol

↓

filtration

↓

centrifugation ⟶ fats, pigments

↓

partition against
ethyl acetate:n-butanol:n-propanol /50:10:5/

DEAE-Sephadex A-25 /acetate/

methanol:n-hexane:water
/100:20:3/

92.6 ml methanol:n-hexane:
H_2O /100:20:3/ + 7.4 ml
acetic acid

↓

ABA-Glc

↓

ABA

hydrolysis with 1N NaOH

Sephadex LH 20
1,2-dichloroethane:methanol
/2:1/

↓

partition against
ethyl acetate /pH 2.5/

↓

Sephadex LH-20
1,2-dichloromethane:methanol
/2:1/

HPLC, GLC

↓

HPLC, GLC

↓

↓

ABA
/released from ABA-Glc/

ABA
/free/

Fig. 3   Purification of ABA and ABA-Glc

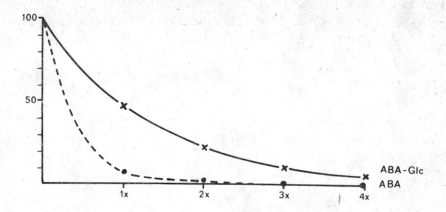

Fig. 4: Extraction of ABA and ABA-Glc by ethyl acetate: n-butanol : n-propanol /50:10:5/. Equal volumes of water solutions and the organic mixture were used.

The separation of ABA and ABA-Glc was performed by ion-exchange chromatography on DEAE-Sephadex A-25 /acetate/. The eluent methanol/n-hexane/water allows the separation of ABA--Glc from other neutral substances /e.g. dyes/ by partition chromatography; consequently, the fractions containing ABA-Glc are colourless. ABA was eluted by addition of acetic acid /7.4 ml/ to the first eluent /92.6 ml/. Gradient elution did not improve the purification of ABA on this column. A second column chromatography was used for further ABA purification. From many columns tested the best version seems to be the partition chromatography on Sephadex LH-20 using 1.2-dichloroethane/methanol /2:1/. After this column the sample was pure enough for the measurement of ABA concentration by HPLC or, after derivations, by GLC. The LH-20 column, calibrated with ABA can be used with high reproducibility for several runs.

For measurement of ABA-Glc the fractions obtained after DEAE-Sephadex chromatography were collected, evaporated and hydrolyzed with 1 N NaOH. Tha ABA yielded was extracted with ethyl acetate and purified on Sephadex LH-20.

For recovery determinations of ABA and ABA-Glc ^{14}C-labelled standards were added to different plant materials /e.g. leaves, needles, roots/ before homogenization. The recovery of ABA was 87-91%, and the recovery of ABA-Glc was 78-83%. The low losses of ABA and ABA-Glc during the separation and purification by column chromatography show clearly the advantage of this method for the measurement of both free ABA and ABA--Glc in plant materials.

REFERENCES

1. Phillips, I.D.J., and Hofmann, A.: Planta 146, 591-596 /1979/

2. Little, C.H.A., Heald, J.K., and Browning, G.: Planta 139, 133-138 /1978/

3. Milborrow, B.V.: J. Exp. Bot. 29, 1059-1066 /1978/

4. Milborrow, B.V., and Vaughan, G.: J. Exp. Bot. 30, 983--995 /1979/

5. King, R.W.: Planta 132, 43-51 /1976/

6. Naumann, R., and Dörffling, K.: Plant Sci. Lett. 27, 111-117 /1982/

7. Dörffling, K., Sonka, B., and Tietz, D.: Planta 121, 57-66 /1974/

8. Dumbroff, E.B., Cohen, D.B. and Webb, D.P.: Physiol. Plant. 45, 211-214 /1979/

9. Hiron, R.W.P., and Wright, S.T.C.: J. Exp. Bot. 24, 769-781 /1973/

10. Terry, P.H., Aung, L.H., and DeHertog, A.A.: Plant Physiol. 70, 1574-1576 /1982/

11. Wright, S.T.C.: J. Exp. Bot. 26, 161-174 /1975/

12. Lehmann, H., Repke, H., Gross, D., and Schütte, H.R.: Z. Chem. 13, 255-256 /1973/

13. Lehmann, H., Miersch, O., and Schütte, H.R.: Z. Chem. 15, 443 /1975/

CAPILLARY LIQUID CHROMATOGRAPHY. PROBLEMS AND PROSPECTS

B.G. BELENKII, E.S. GANKINA, O.I. KURENBIN and
V.G. MALTSEV

Institute of Macromolecular Compounds of the Academy
of Sciences of the USSR, Leningrad, USSR

A decrease in the column diameter in liquid chromatography (LC) is very promising: the consumption of the sorbent and the eluent decreases in accordance with decreasing cross-section of the column, with simultaneous improvement in the minimum detectability of the sample components (assuming that the concentration sensitivity of the detector is retained). There are, however, also other advantages of using small-diameter columns, which are not so evident, such as improvement in the heat transfer and the resulting possibility of increasing the linear elution rate limited by increasing heat evolution caused by friction between the liquid and sorbent particles. This heat evolution deforms the boundaries of the chromatographic zones and thus decreases the column performance. Mass transfer is improved when the column diameter is reduced, and hence it is possible to combine columns into 5-10 m long systems resulting in an efficiency of hundreds of thousands of theoretical plates. Small-diameter columns can also be connected "on line" with a mass spectrometer and other detectors in which large eluent volume is undesirable. A significant reduction in the eluent volume greatly increases the toxic and fire safety of operations of a liquid chromatograph and such "exotic" solvents as trifluoroethanol and trifluoroacetic acid may be used as well as special expensive sorbents.

On the other hand, design and construction of a high-performance liquid chromatograph with small diameter columns involve difficulties mainly due to the necessity of a drastic decrease in extra-column band spreading determined by the volumes of the detector, connecting tubes and injector and a decrease in leaks in the liquid system of the chromatograph. It is also difficult to prepare high-performance small-diameter columns and to ensure high concentration sensitivity of the detectors at a reduced volume of the measuring cell. However, the development of high-performance liquid chromatographs with 1 mm diameter columns (e. g., the Model LC-5 A chromatograph of Shimadzu [1]) shows that these difficulties can be overcome.

It should be emphasized that the chromatography in which small-diameter columns are used should not be regarded as only one of many variations of modern LC. Only with the aid of LC on microcolumns is it possible to attain the record characteristics of the speed, efficiency and sensitivity of analysis.

Let us consider the terminology of liquid chromatography with small-diameter columns. In our opinion, the best expression for this variation of chromatography is capillary liquid chromatography, by analogy with the corresponding variant of gas chromatography.

Capillary liquid chromatography includes two subgroups:

(a) Microbore column liquid chromatography, with packed co-

lumns having a diameter of 0.5-1 mm, packed with 5-10 μm particles.

(b) Microcapillary liquid chromatography with columns having a diameter of 0.1-0.3 mm. These columns can be
packed: packed microcapillary columns or open: open
microcapillary columns .

In the symbols referring to the two column variants we shall
use the acronyms PC for packed columns and OMCC for open microcapillary columns.

The main purpose of analytical LC is the performance of
chromatographic analysis

a) at a maximum flow rate (minimum time),

b) with minimum volume of the chromatographic peak ensuring
the minimum amount of the sample, i. e. the analysis with the
best possible analytical sensitivity and

c) with maximum efficiency making it possible to obtain on
a chromatogram the maximum number of separated components.

Many papers have been published on capillary liquid chromatography. It may be said that microbore-column LC with columns 1 mm in diameter has been completely developed as regards both the instrumentation and the achievement of high
speed and high efficiency of analysis [2,3]. The technique of
the preparation of microbore columns 1.0 mm in diameter with
an HETP equal to two particle diameters $\left(d_p \right)$ has been developed. High-performance liquid chromatographs with microbore
columns, 1 mm in diameter, are manufactured by several companies (e. g. Shimadzu and Jasco [4]). Record sensitivity has
been obtained by using packed ultramicrobore columns 0.2-0.3

mm in diameter, prepared from flexible quartz capillary tubing
[5]. The HETP of these columns approaches two d_p. Some examples have been reported in which superhigh performance has
been obtained by using open microcapillary columns [6]. However, on the whole ultramicrobore and open microcapillary LC

are still in the stage of laboratory development. The difficulties are caused mainly by the construction of detectors
with measuring cells of ultrasmall volume rather than by preparation of high-performance packed or open microcapillary columns. As to packed microcapillary columns we can agree with
Guiochon [7] that such columns with a diameter of 100 and 30
μm developed by Novotny [8] are not very efficient, and the
preparation of packed microcapillary columns with an optimum
particle diameter of 5 μm involves considerable difficulties
and does not provide any advantages over open microcapillary
columns having the same diameter.

Conventional instrumentation for HPLC modified for microcolumns includes a pump with a capacity of less than 10 μl/min,
an injector with a volume 0.2-1 μl and detectors with a measuring cell of 0.5-1 μl. They permit to operate with microbore
columns with a diameter of 1 mm or higher. Such instruments
have been used by most companies. Special instruments are needed for packed microbore columns with a diameter of 0.5 mm.
These instruments are being developed in the USSR [9] and Japan (Jasco [10]).

At present the applications and the main features of common LC and microbore LC are readily distinguished (Table I).
It is not so simple to distinguish between the application

fields of microbore and microcapillary LC because the fields
of application of the latter in high-speed, ultra-sensitive
and super-efficient analysis have not yet been completely determined.

The purpose of the present paper is to consider the efficiency of application of microbore and microcapillary LC to these
types of analyses and the possibility of achieving record
LC parameters and to show the possibility of efficient application of microbore LC to the most complex standard types of
chromatographic analysis: the determination of molecular-weight
distribution of polymers by using exclusion chromatography
and the analysis of amino acids. The possibility of carrying
out these analyses is important for the final establishment
of micro LC as standard chromatographic method.

1. Optimization of liquid chromatography. Comparison
of Microbore and Microcapillary LC.

We will use the reduced Gidding's similarity parameters [11],
such as the reduced HETP (h) and reduced linear elution rate
(ν)

$$h = \frac{H}{d} \tag{1}$$

$$\nu = \frac{ud}{D_m} \tag{2}$$

where H is the HETP; d is the particle diameter in packed micro LC $\left(d = d_p\right)$ and the column diameter in open microcapillary
LC $\left(d = d_c\right)$; u is the linear velocity and D_m is the diffusion
coefficient in the mobile phase.

Knox [12] has determined the relationship between h and ν

for packed columns for a substance with a capacity factor of $k' = 2$.

$$h = 1.8/\sqrt{v} + 1.7v^{1/3} + 0.066v \qquad (3)$$

$$h_{min} = 3.16 , \quad v_{opt} = 2.7 \qquad (4)$$

These values will be assumed to be the most realistic values for the calculation of efficiency in capillary LC, although Scott [3] has shown that in micro LC, it is possible to obtain $h = 2$.

An equation for open microcapillary LC similar to eq. (3) has been obtained by Golay [13]. For $k = 2$ we have

$$h = 2/\sqrt{v} + 0.066v \qquad (5)$$

where

$$h_{min} = 0.78 , \quad v_{opt} = 5.51 \qquad (6)$$

The dependence of $h = h(v)$ for packed (eq. 3) and open microcapillary columns (eq. 5) are shown in Fig. 1. The quality of the chromatographic separation of two substances is determined by Snyder's equation

$$K_R = \frac{1}{4}(\alpha - 1) \ N^{1/2} \ \frac{k'}{1 + k'} \qquad (7)$$

where K_R is peak resolution and N is the efficiency of a chromatographic column (the number of theoretical plates). Good separation is achieved at $K_R \geqslant 1$.

The quality of separation of a complex mixture is determined by the peak capacity, n, the number of chromatographic peaks separated at $K_R = 1$ in a chromatogram:

$$n = 1 + 0.6N^{1/2} \ \log(1 + k'_n) \qquad (8)$$

where k_n is the capacity factor of the last peak in the chromatogram (at $k' = 6.4$ we have $n = 0.5 \ N^{1/2}$).

Eqs. (7) and (8) permit to determine the required efficiency of a separating system. The eluent velocity in a column having the efficiency N may be determined from the parameters of the chromatographic measurement by using Darcy's equation:

$$u = \frac{k_o \, d^2 \, \Delta P}{L \, \eta} \tag{9}$$

or

$$v = \frac{k_o \, d^3 \, \Delta P}{L \, \eta \, D_m} = \frac{k_o \, d^2 \, \Delta P}{N \, h \, \eta \, D_m} \tag{10}$$

where ΔP is the pressure drop in the column, L is the column length ($L = Nhd$), η is the viscosity of the eluent and k_o is the hydrodynamic permeability of the column (for a packed column $k_o = 1/1200$, for an open microcapillary column $k_o = 1/32$ and for a packed microcapillary column $k_o = 1/150$).

Liquid chromatography can be optimized in three ways:

(a) according to the minimum analysis time, $t_{R(min)}$ where

$$t_R = \frac{L}{u} \left(1 + k'\right) = \frac{L \, d}{v \, D_m} \left(1 + k'\right) \tag{11}$$

(b) according to the maximum analytical sensitivity, i.e. the minimum amount of the sample determined by the minimum volume standard deviation, $\delta_{v \, min}$. For a packed column,

$$\delta_v = \frac{L \, \pi \, d_c^2 \, \rho \, (1 + k')}{4 \, N^{1/2}} = \frac{1}{4} \, N^{1/2} \, \pi \, h \, d_p d_c^2 \, \rho \, (1 + k') \tag{12a}$$

and for an open microcapillary column,

$$\delta_v = \frac{L \, \pi \, d_c^2 \, (1 + k')}{4 \, N^{1/2}} = \frac{1}{4} \, N^{1/2} \pi \, h \, d_c^3 \, (1 + k') \tag{12b}$$

(c) according to the maximum efficiency of the analysis:

$$N = \frac{L}{hd} \qquad (13)$$

A. Optimization of the time of analysis.

Proceeding from eqs. (10) and (11) we may write the expression for the time of analysis:

$$t_R = \frac{N^2 h^2 \eta}{k_o \Delta P} \left(1 + k' \right) \qquad (14)$$

Evidently, the decrease in t_r to $t_{R(min)}$ is caused by $h = h_{min}$ and by the increase of ΔP to ΔP_{max}:

$$t_{R\ min} = \frac{N^2 h^2_{min} \eta}{k_o \Delta P_{max}} \left(1 + k' \right) \qquad (15)$$

In this case d (d_p for packed (PC) and d_c for open microcapillary columns (OMCC)) according to eq. (10) depends on ΔP and $h\dot{v}$. Evidently, ΔP_{max} and ($h_{min}\, v_{opt}$) correspond to a certain d_{opt} value:

$$d_{opt} = \left(N \eta D_m \right)^{1/2} \left(\frac{h_{min}\, v_{opt}}{k_o\, \Delta P_{max}} \right)^{1/2} \qquad (16)$$

By using $(h_{min}\, v_{opt})$ PC = 8.5 (Eq.4) and $(h_{min}\, v_{opt})$ OMCC= 4 (eq.6) it is possible to determine the $d_{p(opt)}$ (PC)/$d_{c(opt)}$ (OMCC) ratio:

$$\frac{d_{p(opt)}\ (PC)}{d_{c(opt)}\ (OMCC)} = \left[\frac{(h_{min}\, \dot{v}_{opt})\ PC\ k_o\ OMCC}{(h_{min}\, v_{opt})\ OMCC\ k_o\ PC} \right]^{1/2} \simeq 8.9 \quad (17)$$

Hence, if we assume that d_p(PC)= 5 μm, then d_c(OMCC) 0.6 μm. For technical reason it is difficult to utilize such small diameters and, as can be seen from further considerations, it will require a detector cell of very small volume. If the condition d_c (OMCC) = d_p (PC) is accepted, then according to eq.(16) $(h v)_{OMCC}$ = 318.8. Then applying eq.(5) we have $v =$

$[(h \vee - 2)/0.066]^{1/2} = 69.3$ and $h = 4.60$.

It is of interest to compare the t_r PC $/t_r$ OMCC ratio at d_{opt}, i.e. when $h = h_{min}$ and at $d_p(PC) = d_c(OMCC)$:

$$\frac{t_R(PC)}{t_R(OMCC)} = \frac{h_{min}^2(PC)}{h_{min}^2(OMCC)} \frac{k_o \text{ OMCC}}{k_o \text{ PC}} \approx 704 \qquad (18)$$

$$\frac{t_R(PC)}{t_R(OMCC)} = \frac{h_{min}^2(PC)}{h^2(OMCC)} \frac{k_o \text{ OMCC}}{k_o \text{ PC}} = 17.5 \qquad (19)$$

When d_c undergoes further increase by a factor of two, the $t_R(PC)/t_R(OMCC)$ ratio decreases to 4.3. Hence, the increase in $d_c(OMCC) > d_{c(opt)}$ results in losing the advantage of an open microcapillary column with respect to the rate of analysis.

These relationships are shown in Fig. 2 in which the dependencies of t_R on ΔP at different d values are plotted for packed and open microcapillary columns. It is clear that as ΔP increases, the dependencies $t_R = t_R(\Delta P)$ reach a plateau and with increasing d this is observed at lower ΔP values. If the values of ΔP and d for the two types of columns are equal, then the value of t_R for the open microcapillary column is lower than for the packed column and the dependence $t_R = t_R(\Delta P)$ for the former reaches a plateau at lower ΔP values. The reason for these phenomena is the fact that $\vee$ increases with ΔP, h becomes a linear function of $\vee$ and hence, according to eq.(10) h^2 becomes proportional to ΔP. Then according to eq.(14) t_r no longer depends on ΔP. Evidently, the higher the k_o and d values, the lower the ΔP values at which the dependence $t_R = t_R(\Delta P)$ reaches a plateau.

Let us now compare the time of analysis for packed and open microcapillary columns when d is fixed and ΔP is not limited:

$$t_R = \frac{N_d^2}{D_m} \left(1 + k'\right) \frac{h}{v} \tag{20}$$

As ΔP increases, h/v decreases at first but when it reaches the value of h_c/v_c t_R no longer depends on ΔP (dependences in Fig. 2 attain a plateau). For a packed column, we have $v_c = 70$, $h_c = 10.5$, $c = 0.15$ and for an open microcapillary column we have $v_c = 30$, $h_c = 2.04$, $c = 0.68$. Under these conditions, at $d_p(PC) = d_c(OMCC)$ we have

$$\frac{t_{R,c}(PC)}{t_{R\,c}(OMCC)} = \frac{(h_c/v_c)_{PC}}{(h_c/v_c)_{OMCC}} = 2.21 \tag{21}$$

i.e. $t_{R,c}(PC)$ and $t_{R,c}(OMCC)$ are virtually equal. In this case according to eq.(10) the ratio of the pressure required for packed and open microcapillary columns is given by

$$-\frac{\Delta P_c(PC)}{P_c(OMCC)} = \frac{(h_c v_c)_{PC}}{(h_c v_c)_{OMCC}} \frac{k_{o\ OMCC}}{k_{o\ PC}} \simeq 450 \tag{22}$$

Hence, it becomes evident that chromatography with an open microcapillary column has considerable advantages over packed columns with respect to the time of analysis at equal P and with respect to the pressure required at equal t_R.

The most important chromatographic parameters considered here are ΔP and, as will be seen below, the volume of the detector measuring cell, V_d. At fixed ΔP $t_R = t_{R\ min}$ is obtained at $h = h_{min}$ and $d = d_{opt}$ is difficult because it is difficult to prepare such a column with $d_c < 1\ \mu m$ and to manufacture a detector of ultrasmall diameter.

In this connection the effect of column diameter on the volume of detector measuring cell will be considered. It is assumed that this volume, V_d should be more or less equal to one half of the standard deviation of the peak of an unretained component, δ_v^o

$$V_{d\,(PC)} = \frac{\delta_v^o}{2} = \frac{\pi}{8} N^{1/2} h\, d_p\, d_c^2\, \beta = 0.41\, N^{1/2}\, d_p\, d_c^2 \quad (23a)$$

$$V_{d\,(OMCC)} = \frac{\delta_v^o}{2} = \frac{1}{8} N^{1/2} h\, d_c^3 = 0.28\, N^{1/2}\, d_c^3 \quad (23b)$$

In eq. (23a) and (23b) it is assumed that $h = h_{min}$. If $d_{c\,(opt)} = 0.6\,\mu m$, then according to eq. (23b) $d_{OMCC} = 6 \cdot 10^{-14}$ $N^{1/2}(cm^3)$ and even for a column with $N = 10^7$ we have $V_d = 2 \times 10^{-4}$ nl, i.e., the value of V_d is unreal. Let us increase $d_{c\,(OMCC)}$ to $10\,\mu m$. Then the $V_{d\,(OMCC)}$ value for columns exhibiting various efficiencies should be

N_{OMCC}	10^4	10^5	10^6	10^7
$V_{d(OMCC)}$(nl)	$2.8\ 10^{-2}$	$8.9\ 10^{-2}$	0.28	0.89

At present detectors for microcapillary liquid chromatography with $V_d \simeq 1$ nl are known such as the flame-ionization [14], flame-emission [14] and electrochemical detectors [15] and, presumably, mass spectrometry [16]. For a laser fluorimetric detector with a hydrodynamic cell the cell volume is a few nanoliters [17]. Hence, a detector open microcapillary columns with $N = 10^7$ is available. This is, however, not the case for open microcapillary columns with $N < 10^6$.

Let us now consider the problem of the detector for the packed columns. Since the diameter is not critical for the efficiency of packed columns, it may be selected proceeding from the volume of the measuring cell of the available detector. According to eq. 23a , we have

$$d_{c\ PC} = \left(\frac{V_{d(PC)}}{0.41\ N^{1/2}\ d_p} \right)^{1/2} \tag{24}$$

Eq. (24) shows that $d_{c\ PC}$ depends not only on $V_{d\ PC}$ but also on column performance and the diameter of sorbent grains.

At present a number of detectors with $V_d \simeq 1\ \mu l$ are known and therefore it is desirable by using eq. (24) to determine $d_{c\ PC}$ for capillary microchromatography with various efficiencies which a detector with $V_d = 1\ \mu l$ may ensure when particles with $d_p = 5$ and $10\ \mu m$ are used.

N_{PC}		10^4		10^5		10^6	
$d_p(\mu m)$	5		10	5		10	5	10
d_c (mm)	2.24		1.58	1.26		0.89	0.71	0.495

Hence, a detector with $V_d = 1\ \mu l$ is too large for packed columns with $d_c = 0.5 - 1$ mm generally used in microcolumn chromatography. For packed columns with $N = 10^4$ the value of $d_c = 1$ mm requires a detector with $V_d = 0.2\ \mu l$ for a sorbent with $d_p = 5\ \mu m$ and $V_d = 0.4\ \mu l$ for sorbent with $d_p = 10\ \mu m$.

As already shown in eqs. (21) and (22), at approximately equal analysis times packed columns require a much higher pres-

sure than open microcapillary columns (about 200 times higher).
The highest allowable pressure, ΔP_{max}, limits both the minimum
time of analysis at given N and the performance of the column,
particularly that of a packed column, at a given t_R value. As
indicated by Guichon [7], the value of ΔP_{max} is related not
only to the technical possibilities of the pump and the injection pressure for which the limits are probably 300 - 500 MPa
and 100 MPa, respectively. At higher pressures it is possible
to use injection valves with stop-flow [7]. However, the increase in the pressure to values higher than 100 MPa leads to
a change in the viscosity of the solvent, the diffusion coefficient and sorption constants [7]. Some solvents, such as
cyclohexane, at $\Delta P = 300$ MPa freeze at room temperature. However, most solvents including acetonitrile, methanol and water remain liquid at this pressure. Another negative effect
of high pressure is due to the heating of the liquid as it
flows at a high velocity through a packed column [18]. This
effect leading to the distortion of zone boundaries and, cor-
respondingly, to an increase in the HETP, decreases with increasing heat transfer with a decrease in d_c. Hence, it is
possible to carry out super high-speed chromatography on
packed microcolumns.

B. Ultrasensitive analysis

For ultrasensitive analysis a minimum standard deviation
δ_v should be attained because the substance concentration at
peak maximum determining the minimum amount of the analyte is
controlled by δ_v:

$$q^* = c^* \sqrt{2\pi} \, \delta_v \tag{25}$$

where q^* is the minimum amount of the analyte and C is the minimum concentration analyzed.

By using the equation for σ_v (eq. 12) we can express the value of q^* for packed and open microcapillary columns as follows:

for a packed column:

$$q = \frac{\pi\sqrt{2\pi}}{4}\,C^*N^{1/2}hd_p\beta\,d_c^2\left(1 + k'\right) = 0.66 C^*N^{1/2}hd_p\beta\,d_c^2\left(1+k'\right) \quad (26a)$$

for an open microcapillary column:

$$q = \frac{\pi\sqrt{2\pi}}{4}\,C^*N^{1/2}hd_c^3\left(1+k'\right) = 1.97\,N^{1/2}hd_c^3\left(1+k'\right) \quad (26b)$$

It can be seen from eqs. (26a) and (26b) that the minimum amount of the analyte depends on h and d_c, in the case of a packed column it also depends on d_p (at fixed N and C^* values). Evidently, the smaller the column diameter, the smaller is the amount of the sample, which is proportional to d_c^2. The absolute minimum of q^* depends on the chromatographic conditions when $h = h_{min}$.

For high-speed analysis, when ΔP is not limited (eq. 22) and $h \to h_c$, the sensitivity of the analysis decreases in accordance with the h_{min}/h_c ratio (for a packed column by a factor of 2.37 and for an open microcapillary column by a factor of 2.83). As shown above, the effect of the column diameter on q decreasing this value is limited by the volume of the detector measuring cell $0.2 - 1\ \mu l$ for a packed column and ~ 1 nl for an open microcapillary column. Hence, the use of open microcapillary columns for ultrasensitive analysis is advisable only at $N \geqslant 10^6$. However, the possibility of carrying out the analysis at $N \geqslant 10^6$ may limit the sensitivity of the detector.

In the investigation of the problems related to ultrasensitive analysis it is of interest to determine the dependence of σ_v on t_R for a column of a certain length, L. This dependence becomes apparent in the effect of v on h (Fig. 1). Fig. 3 shows that the lower t_R, i.e., the higher v, the greater is σ_v. In this case the increase in the speed of analysis in a certain range of d leads to a decrease in the value of σ_v. This phenomenon will be observed if the value of h approaches h_{min} from the left, from the side of low v.

C. Economic indices of chromatographic analysis

The consumption of the sorbent and the eluent decreases with the column diameter. Naturally this method of improving the economic indices of the chromatographic analysis is promising only for packed columns. The column volume, V_c, and the eluent volume, V_R, depend on d_c^2

$$V_c = \frac{1}{4}\pi d_c^2 L = \frac{\pi}{4} N d_p h d_c^2 \tag{27}$$

$$V_R = \frac{1}{4}\pi d_c^2 \beta L^{(1+k')} = \frac{\pi}{4} N d_p h \beta (1 + k') d_c^2 \tag{28}$$

The following points should be considered here:

(a) The decrease in the elution eluent volume not only improves the economic indices of the chromatograph but also drastically decreases the toxic and fire hazards of its operation. It is also possible to use expensive and toxic solvents and special expensive sorbents.

(b) The use of microbore columns presents specific problems because a leakage of solvents from the liquid system of the chromatograph is possible and it is necessary to ensure the

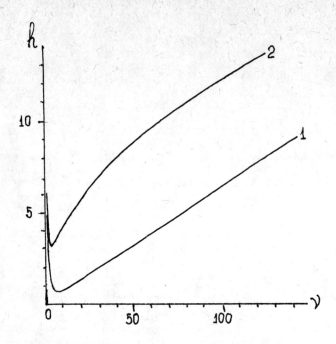

Fig. 1. Dependence of h on ν for 1) open microcapillary columns; $h = 2/\nu + 0.066\nu$ and 2) packed columns: h =
$1.8/\nu + 1.7\nu^{1/3} + 0.05\nu$ at $k' = 2$.

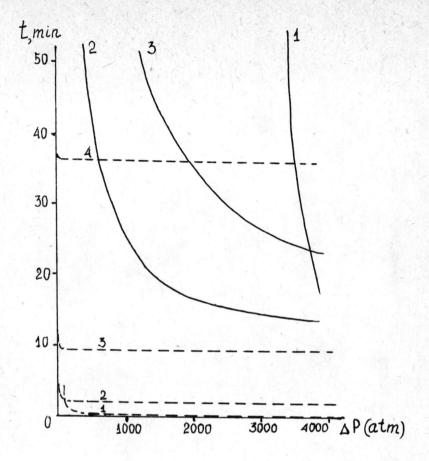

Fig. 2. Dependence of t_R on ΔP for packed (———) and open
microcapillary (-----) columns at d = 1) 1 μm, 2) 5
μm, 3) 10 μm and 4) 20 μm; N = 10^5, β = 1/3, η = 0.4
cP, D_m = 4 10^{-6} cm^2/s, k' = 2, k_o = 1/1200 (PC), 1/32
(OMCC), t_R = $Nhd^2(1 + k')/\sqrt{}$ D_m, h $\sqrt{}$ = $\kappa_o d^2 \Delta P/N\eta D_m$

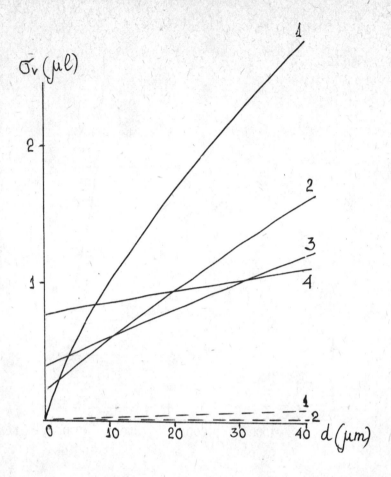

Fig. 3. Dependence of σ_v on d for packed (———) and open microcapillary (-----) columns at t_R = 1) 1 min, 2) 10 min, 3) 30 min and 4) 120 min; L = 30 cm, $\sigma_v = (\pi/4)\left(Ld_p/h\right)^{1/2}\beta\, d_c^2(1 + k') - PC$, $\sigma_v = (\pi/4)(L/d_c^5 h)^{1/2}(1 + k') - OMCC$, h = h($\nu$), ν = Ld $(1 + k')/t_R D_m$, d_c PC = 0.5 mm, β = 1/3.

reproducibility and precision of eluent introduction at low eluent consumption (<10 μl/min for packed columns). The danger of leakage increases for microbore columns since the probability of leakage in the fittings is proportional to the circumference of the column, to d_c and the negative effect of leakage is due to imprecise volume maintenance, i.e. it is proportional to d_c^2. Hence, with decreasing d_c the reproducibility of volume introduction of the eluent decreases with $1/d_c$. Reciprocating pumps with eluent consumption of less than 10 μl/min at 3% reproducibility have recently been developed (Shimadzu LC-5A chromatograph). These chromatographs may be used with packed microbore columns with a diameter of 1 mm or higher. For smaller-diameter microbore columns syringe-type pumps are used. The reproducibility and precision of their operation are ensured by the precision of the manufacture of the lead screw and by the low ratio of the cross sectional area of the column to that of the pump plunger. For this reason the Familic 100 N chromatograph uses a Hamilton microsyringe. The higher is this ratio, the higher are the precision and reproducibility of the micropump. However, there are technical limits to the decrease in the diameter of the pump plunger because the sealing of the pump and the manufacture of a long lead screw involve difficulties.

(c) In capillary liquid chromatography the extracolumn band spreading must be significantly decreased, mainly the volume of the detector measuring cell V_d, and this decrease leads to inferior detector characteristics. In the case of photometric detector this results in a decrease of the optical length of the cell and a decrease of the light flux passing through it and hence leads to increasing noise which is inversely proportional to the square root of the intensity of the light flux. A decrease in cell length decreases the concentration sensitivity of the interference, electrochemical and fluorimetric detectors. The only detector unaffected by the volume of the measuring cell is the refractometric detector in which the change in the refraction index of the elute is related to optical phenomena occurring at the liquid-glass interface. In this case it is difficult to prepare a microcell of desired dimensions and shape. These difficulties can be overcome more easily in a fluorimetric detector because its sensitivity can increase almost infinitely when the optical noise caused by the Raman and Raleigh scattered light and by the fluorescence of the solvent and optical elements decreases. Great possibilities for improvement are provided by laser fluorimetry [19] which makes it possible to avoid the presence of scattered light without difficulty.

Unfortunately, the greatest losses in sensitivity are caused by the decrease in the cell volume of the photometric (spectrophotometric) detector which is of major importance in LC. Probably the only solution to this problem is the application of a laser with variable frequency as the radiation source in which the radiation density is $6 \cdot 10^3 - 10^4$ higher than in common light sources. Moreover, it is possible to manufacture a long measuring cell of small diameter and hence having a small volume.

As to the detectors for ultramicrobore LC and microcapillary LC, it is still necessary to use the detectors the concentration sensitivity of which is **much** lower than the sensitivity of detectors for micro-LC and, to even greater extent, than the sensitivity of the detectors used in HPLC with conventional columns. Hence, it can be stated that unfortunately the increase in column diameter has not yet resulted in a decrease in the minimum sample amount, q^*, that might be expected if as a high concentration sensitivity of the detector is combined with a very small cell volume.

(d) A decreasing column diameter involves manufacturing difficulties. Thus, when metal packed microbore columns with a diameter of 1 mm and h = 2 are prepared, a pressure of up to 250 MPa should be used [2], whereas for conventional columns a pressure of 50-60 MPa is sufficient. On the other hand, by using flexible quartz capillaries it is possible to prepare packed ultramicrobore columns with d_c = 0.2 mm and h $\simeq$ 2 at a pressure of 50-60 MPa [5]. Unfortunately, the possibilities to prepare high-performance open microcapillary columns with d_c = 5-10 μm have not yet been investigated in detail.

D. Super-efficient chromatographic analysis

Improvement of mass transfer in small-diameter columns makes it possible to develop super-efficient chromatographic systems (with $N > 10^5$). In this case the problem lies in the development of systems having a maximum efficiency which could separate the maximum number of components (i.e., would have the maximum peak capacity). These systems would probably be useful mainly in the petroleum chemistry and in the separation of isotope-labeled compounds and complex natural samples. Both

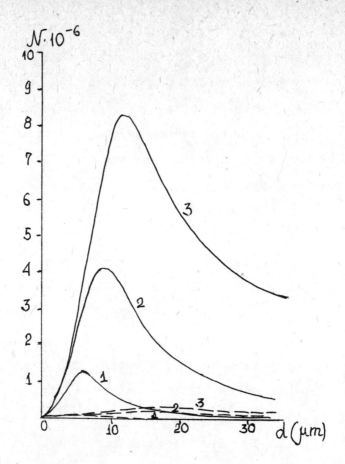

Fig. 4/a. Dependence of N on d ΔP for packed (————) and open microcapillary (-----) columns at ΔP = 10 MPa; t_R= 1) 1h, 2) 10h, and 3) 40h; $N = \left[t_R \Delta P k_o / h^2 \eta (1 + k) \right]^{1/2}$, $h \sqrt{} = k_o d^2 \Delta P / N \eta D_m$

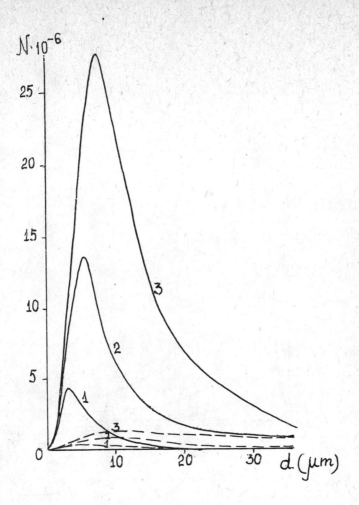

Fig. 4/b. Dependence of N on d ⚠ P for packed (———) and open
microcapillary (-----) columns at ⚠ P = 100 MPa; for
the other parameters see Fig. 4/a.

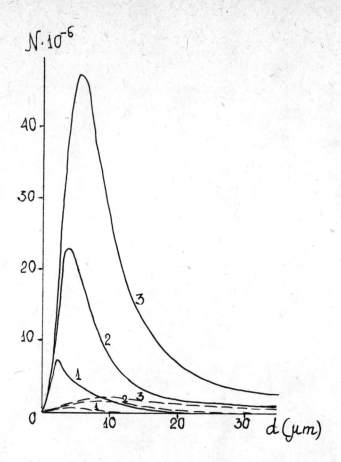

Fig. 4/c. Dependence of N on d ΔP for packed (———) and open
microcapillary (------) columns at ΔP = 300 MPa; for
the other parameters see Fig. 4/a.

packed microbore and open microcapillary columns may be successfully used in these systems because the permissible volume
of the measuring cell increases with N. The only limitations
on the development of super-efficient separating systems are
the maximum allowable ΔP_{max} and the time of analysis. It follows from eq. (14) that two cases are possible here:

(a) optimum conditions when $h = h_{min}$. Then in accordance
with eq. (14) the column performance is given by

$$N = \frac{1}{h}\left[\frac{t_R\, k_o\, \Delta P}{\eta\,(1 + k')}\right]^{1/2} \tag{29}$$

(b) conditions of high-speed analysis, where $h > h_{min}$
($h \to h'$), under which N may be calculated according to the equation derived from eq. (20):

$$N = \frac{t_R\, D_m}{d^2\left(1 + k'\right)\left(h/\nu\right)} \tag{30}$$

In the first case N is proportional to ΔP at a fixed time
of analysis, while in the second case, N is limited by the
condition $h/\nu = h'/\nu' = C$ which is reached at ΔP_{lim}:

$$\Delta P_{lim} = \frac{\left(h'\nu'\right)N\, D_m\, \eta}{k_o\, d^2} \tag{31}$$

Above this value the increase in ΔP does not lead to an
increase in N. In this case, as can be seen from eq. (31), the
value of ΔP_{lim} decreases with increasing d. In order to use
eqs. (29) and (30) it is necessary to determine ν as a function of t_R, ΔP and d. Applying eqs. (10) and (11) we obtain
the following expression:

$$\nu = \frac{d^2}{D_m}\left[\frac{k_o\, \Delta P\,(1 + k')}{t_R\, \eta}\right]^{1/2} \tag{32}$$

At a fixed ΔP value under the optimum analytical conditions $\left(h = h_{min} \right)$

$$\frac{N_{max}(\text{OMCC})}{N_{max}(\text{PC})} = \frac{h_{min}(\text{PC})}{h_{min}(\text{OMCC})} \left(\frac{k_o\ \text{OMCC}}{k_o\ \text{PC}} \right)^{1/2} = 26.5 \qquad (33)$$

the performance that can be attained obtained in open micro-capillary columns is higher by a factor of 26.5 than for packed columns. In this case, according to eq. $\left(16 \right)$, the value of N_{max} will be obtained at the following ratio of $d_p(\text{PC})$ to $d_c(\text{OMCC})$

$$\frac{d_p(\text{PC})}{d_c(\text{OMCC})} = \left[\frac{N_{max}(\text{PC})\,(h\sqrt{})_{\text{PC}}\ k_o\ \text{OMCC}}{N_{max}(\text{OMCC})\,(h\sqrt{})_{\text{OMCC}}\ k_{o\text{PC}}} \right]^{1/2} = 1.73 \qquad (34)$$

Hence, at virtually equal d values, eq. $\left(34 \right)$ should be compared with eq. $\left(17 \right)$, the ratio of $d_p(\text{PC})$ to $d_c(\text{OMCC})$, under the conditions of high-speed analysis.

Fig. 5 shows the dependences $N = N(d)$ at various ΔP values according to eq. $\left(29 \right)$. The maximum of these dependences, $N = N_{max}$, is observed at $h = h_{min}$. This point may be obtained when $\sqrt{}$ increases with increasing d at a fixed ΔP value. The point $h = h_{min}$ is related to the point $d = d_{opt}$ which is determined as follows:

$$d_{opt} = 2.34 \left[\frac{t_R\ \eta\ D_m}{k_o\ \Delta P\ (1+k')} \right]^{1/4} \qquad (35)$$

It is clear from Fig. 5 and eqs. (29) and (35) that N_{max} increases with increasing t_R and ΔP, whereas d_{opt} increases with increasing t_R and decreasing ΔP.

Fig. 6 shows the dependences of N on ΔP for a packed and open microcapillary column at a fixed t_R $\left(48\ h \right)$ and different d under the conditions of high-speed analysis, i.e., calculated with the aid of eqs. (30) and (32). Under these conditions $\Delta P = 100$ MPa ensures the performance of a packed column with $d_p = 10\ \mu m$ of about 1 million theoretical plates and that of an open microcapillary column with $d_c = 10\ \mu m$ of approximately 25 millions theoretical plates. With increasing d the initial slope of the $N = N(\Delta P)$ curves increases but they more rapidly reach the plateau at $\Delta P = \Delta P_{lim}$ and $h = h'$ at a limited column length $L = Nh'd$. When the length is not limited optimum analytical conditions are ensured $(h = h_{min})$.

The use of super high-performance open microcapillary columns is limited by the volume of the measuring cell of the detector, V_d. It is clear from eq. (23) that the allowable value of V_d depends on $N^{1/2}$, and, therefore, the value of $N = N^+$ corresponding to V_d should be determined above which it is possible to use open microcapillary columns for super-efficient analysis. The substitution of the expression for h from eq. (29) into eq. (23b) gives the equation for N^+ at fixed values of $t_R = t_R^*$ and $\Delta P = \Delta P^*$

$$N > 0.005\ \frac{t_R^*\ \Delta P^*}{\eta(1+k')}\ \frac{d_c^{\,6}}{V_d^{\,2}} \tag{36}$$

In this case the possibility of maintaining a constant value of N^+ at fixed values of t_R^* and ΔP^* (η and k') is determined by the constancy of the $d_c^{\,6}/V_d^{\,2}$ ratio.

Evidently, the increase in $V_d^{\,2}$ requires the corresponding increase in $d_c^{\,6}$. This, in turn, will lead to increasing column length and increasing δ_v. As a result, the sensitivity of the analysis will decrease (q^* will decrease).

It should be noted that the N and t_R values contained in

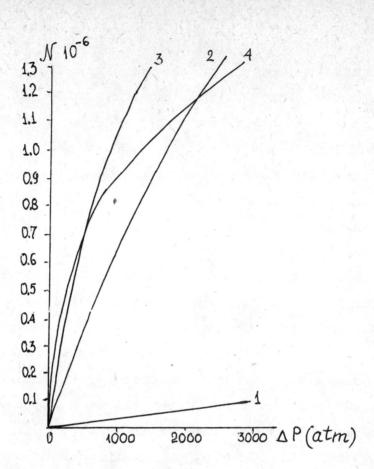

Fig. 5/a. Dependence of N on ΔP for a packed column; d =
1) 1 μm, 2) 5 μm, 3) 10 μm and 4) 20 μm; t_R = 48 h

$N = t_R D_m / d^2 (1 + k') (h/v)$, $hv = k_o d^2 \Delta P / N \eta D_m$

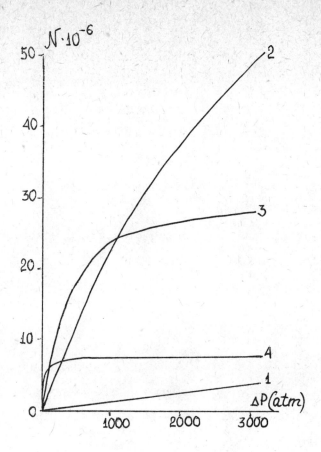

Fig. 5/b. Dependence of N on ΔP for an open microcapillary
column; for the other parameters see Fig. 5/a.

the inequality expressed in eq. (36) refer to the peak of the unretained substance and hence we shall denote them by N_o and t_{Ro}. The elution time of the i-th component (i) is $t_R = t_{Ro} \times (1 + k_i')$. It is noteworthy that V_d and d_c in eq. (36) are inter-related. Their ratio is determined by the values of t_{Ro}, P and N_o $\left(V_d/d_c^3\right)^2 = 0.005\ t_{Ro}\ \Delta P/N_o\ \eta$. The system becomes op-timum only when this equation is obeyed and it is possible to attain maximum efficiency at given values of t_{Ri}, ΔP and N_o.

Let us obtain the dependence of V_d on N according to eq. (36). Applying eq. (14) we can find the value of h_o $(k'=0)$ at given values of t_{Ro} and N_o and then, by using the h_o value and Taylor's equation $(k'= 0)$ $h = 2/\sqrt{} + 0.0104\sqrt{}$, the values of γ_1 and γ_2 can be determined and the values of d_{c1}, and d_{c2} can be obtained with the help of Darcy's equation (eq. 9). The va-lue of V_d is determined from eq. (36). This dependence of V_d on N_o is shown in Fig. 6a. It is clear that the higher ΔP, the greater the value of N beginning from which it is possib-le to use open microcapillary columns. In this case, with in-creasing ΔP, the peak of the maximum of the dependence $N = N(V_d)$ becomes sharper, and the requirement that optimum con-ditions $\left(d_{c\ (opt)}\ \text{and}\ V_{d\ (opt)}\right)$ (when $h = h_{min} \neq 0.288$) should be obeyed becomes more stringent. The dependences shown in Fig. 6 are interesting not only because they determine the possibi-lity of using open microcapillary columns with a detector having a predetermined volume of the measuring cell but main-ly because they determine the condition for the optimization of OMCC with respect of N taking into account extra-column band spreading (V_d) and permitting to reach very high efficiencies. For $t_R = 16$ h these optimum parameters for open microcapillary columns have quite real values: at $\Delta P = 10$ MPa $d_{c(opt)} = 20.3 \mu m$, $V_d = 4.6$ nl and $L = 136$ m and at $\Delta P = 100$ MPa $d_{c(opt)} = 11.5 \mu m$, $V_{d(opt)} = 1.47$ nl and $L = 244$ m. They ensure $N_{max} = 2.3 \cdot 10^7$ and $7.4 \cdot 10^7$ theoretical plates at $\Delta P = 10$ and 100 MPa, respectively, and these values of N_{max} make possible to obtain a peak capacity of $n = 2400$ and 4180, respectively for $t_R = 118$ h $\left(k_i = 6.4 \right)$. A decrease in t_{Ro} leads to a decrease in the values of N_{max}, $d_{c\,opt}$ and $V_{d\,opt}$, and V_d becomes too low for real experiment $\left(\text{Fig. 6b} \right)$. These results show a real possibility of using open microcapillary chromatography beginning from $N = 1$ million theoretical plates However, up to $N = 5 \cdot 10^5 - 10^6$ it is advisable to use microbore columns for which the value of V_d is not critical and N is limited by ΔP_{lim} $\left(\text{at equal } N \text{ the value of } \Delta P_{lim} \text{ for a microbo-} \right.$ re column is higher by a factor of 37.5 than for an open microcapillary column$\left. \right)$. This conclusion, that it is possible and desirable to use open microcapillary columns beginning from $N = 10^6$ is also valid in the case of ultrasensitive superhigh-speed analysis.

The main trend in capillary liquid chromatography is improvement in the method with respect to the sensitivity of the analysis and the decrease in solvent consumption.

Excellent results regarding sensitivity $\left(c^* \text{ in the range} \right.$ of 1 - 10 pg$\left. \right)$ have been obtained for DNS amino acids [20] isomeric amino phenols [21] and amino acids and peptides [22], by using detectors with high concentration sensitivity $\left(\text{fluo-} \right.$ rimetric or electrochemical detector$\left. \right)$. Concerning supereffi-

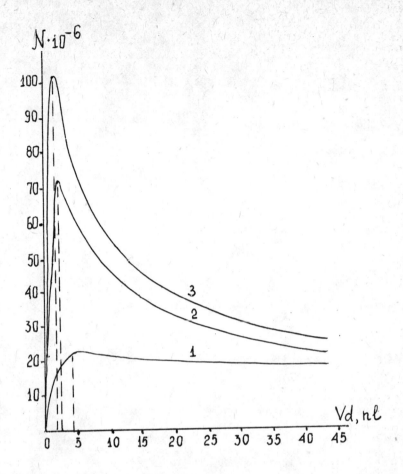

Fig. 6/a. Dependence of N on V_d for an open microcapillary column at $t_{Ro} = 16$ h; $\Delta P =$ 1) 10 MPa, 2) 100 MPa and 3) 200 MPa. At N_{max} the following values exist:

1) $d_c = 20.3 \, \mu m$, $L = 136$ m

2) $d_c = 11.5 \, \mu m$, $L = 244$ m

3) $d_c = 9.6 \, \mu m$, $L = 283$ m

$N \geqslant 0.005 \left(t_{Ro} \Delta P / \eta \right) \left(d_c^6 / V_d^2 \right)$; $\eta = 0.4$ cP, $k_o = 1/32$

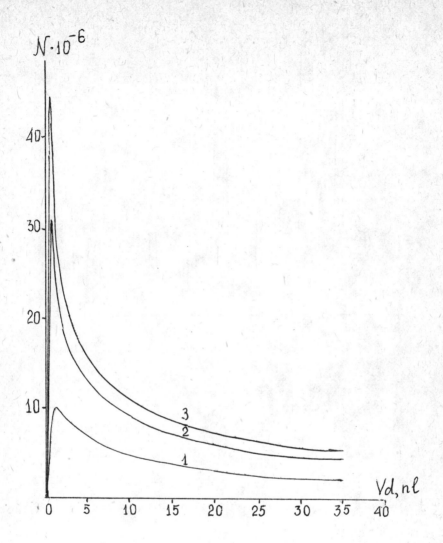

Fig. 6/b. Dependence of N on V_d for an open microcapillary co-
    lumn at t_{Ro}= 3 h; ΔP = 1) 10 MPa, 2) 100 MPa, and
    3) 200 MPa. At N_{max} the following values exist:
    1) d_c = 13.4 μm, L = 39 m
    2) d_c = 7.6 μm, L = 70 m
    3) d_c = 6.4 μm, L = 83 m
    Other parameters as in Fig. 6/a

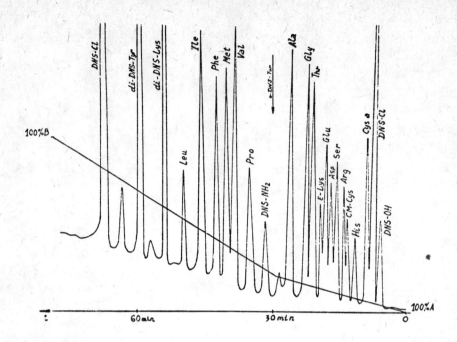

Fig. 7. Microbore column LC of a standard dansylamino acid mix-
        ture on a column 250 mm x 0.5 mm ID packed with sila-
        sorb 300 - C_{18}, d_p = 7 μm. Flow rate 1 ml/h. Eluents:
        A - acetonitrile (25%) - 0.01 M sodium formate, B -
        acetonitrile (60%) - 0.01 Na_3PO_4, pH = 7.0. Sample
        $2.5 \cdot 10^{-13}$ mole of each amino acid. Fluorimetric detec-
        tion.

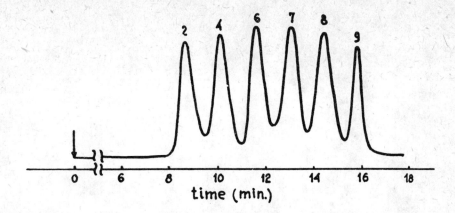

Fig. 8. Chromatogram of polystyrene standards. Column 300 x
0.6 mm packed with a mixture of Lichrospher SI-100 and
SI-1000 (2 : 3). Eluent: methylene chloride, flow rate
4.6 μl/min. Sample volume 0.5 1 5 mg/ml . Detection
at λ = 260 nm. M_w: $2.61 \cdot 10^6$(1) ; $8.67 \cdot 10^5$ (2); $4.11 \cdot 10^5$
(3) ; $2.0 \cdot 10^5$ (4); $1.11 \cdot 10^5$ (5); $3.3 \cdot 10^4$ (6); 10^4 (7);
$2.1 \cdot 10^3$ (8); 78 (9) (benzene) .

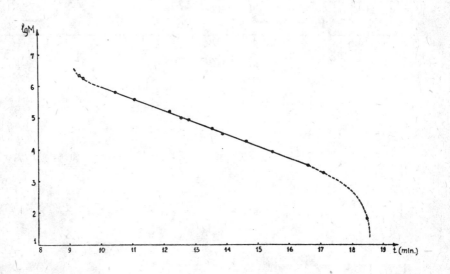

Fig. 9. Retention times vs. logarithm of the molecular weight
of PS standards. Column 330 x 0.6 mm packed with a
mixture of LiChrospher Si 100 and Si 1000 sorbents.
Eluent: methylene chloride, flow rate 4.6 μl/min. Sample volume 0.3 μl (1 mg/ml).

cient or superhigh-speed analysis, Scott and Kucera have attained for microbore columns $6.5 \cdot 10^5$ theoretical plates [23] and $N/t = 280$ [3]. Maximum results for the performance of open microcapillary columns have also been described, obtaining $1.25 \cdot 10^6$ theoretical plates [24]. It should be noted that many investigations of capillary liquid chromatography have been carried out with the only purpose to demonstrate the advantages of this method and thus no conclusions could be made with respect to the applicability of the method to standard quantitative analysis.

In order to carry out in the practice quantitative analysis of complex mixtures by capillary liquid chromatography, it is necessary to ensure the reproducibility of sample injection as well as of the elution volume and peak height in gradient elution, and in exclusion chromatography of polymers it is also necessary to carry out linear calibration over a wide molecular weight range.

These problems have been solved [20] dealing with quantitative analysis (to within 3%) of DNS amino acids in the amount of 10^{-13} mol under the conditions of gradient elution (Fig. 8) see also [25] describing the application of exlusion microbore LC to the determination of the molecular weight distribution of polymers (Figs. 8, 9). In this case the separation efficiency of narrow-disperse standards on a 30 cm long column was not inferior to that on a system of containing 150 cm long columns packed with Lichrospher Merck sorbent at approximately equal analysis times (16 and 12 min, respectively). The determined molecular weight dispersion of a standard polystyrene sample, PS-706, was in good agreement with the specified data (26).

	exclusion LC	[26]
M_n	$(148 \pm 5) \cdot 10^3$	$136.5 \; 10^3$
M_w	$(273 \pm 5) \cdot 10^3$	$(278 \pm 5) \cdot 10^3$

Further prospects of the development of capillary LC

At present it may be considered as proven that capillary LC is not only an ultra-sensitive, high-performance and economical, but also a relatively precise method for the analysis of complex samples. Hence, it might be expected that the part played by microbore column chromatography will continuously increase, particularly for carrying out ultra-sensitive analysis (analysis of trace amounts), and large-scale analytical work involving the consumption of expensive solvents or the analysis of mixtures with very complex composition.

Table II gives an approximate evaluation of the ratio of conventional and microbore column chromatography used in different types of laboratories. The real application field of microcolumn and open microcolumn LC is superhigh performance chromatography with $N > 5 \cdot 10^5$ theoretical plates for the investigation of very complex natural samples. In this case it is desirable to combine microcapillary LC with a very sensitive mass spectrometer. This system is also useful for the isotopic analysis of organic substances labeled with deuterium and tritium. As to microcapillary LC using packed microcapillary columns, one should agree with Guiochon [7] that these are not very promising.

Table I. Comparison of conventional and microbore column liquid chromatography

	Conventional chromatography	Microbore column chromatography
1. Application		
a) Determination of solution concentration	++	+
b) Analysis of microsamples	+	+++
c) Analysis of trace amounts	+	++
d) Micropreparative chromatography	++	+
e) Preparative chromatography	++	-
f) Separation of complex mixtures	++	++
g) Super high-speed analysis	+	++
h) Superefficient analysis	+	+++
i) Precision chromatography	++	+
2. Economic indexes		
a) Cost of the sorbent and the eluent	+	+++
b) Development of analytical procedures with very expensive sorbents and eluents	+	+++
c) Toxic and fire safety	+	+++
d) Equipment cost		
Pumps	+	++
Detectors	+	+
Columns	+	++
Computers	+	+

3. Technical possibilities
   a) Eluent and flow programming   ++   +
   b) Automatic sample injection   +   -
   c) Multidimensional chromatography   ++   +
   d) Number of eluent components   ++   +
   e) High-temperature chromatography   +   +
   f) Automatic search for optimum condi-
      tions   ++   +
   g) Automatic fraction collection   ++   +
4. State of instrumentation development   ++   +
5. Availability of detectors
   a) Photometer   ++   +
   b) Spectrophotometer   ++   +
   c) Fluorimeter   +   +
   d) Refractometer   +   +
   e) Electrochemical detector   +   +
   f) Chemical detector   ++   +
6. Types of chromatography
   a) Exclusion chromatography   ++   +
   b) Adsorption chromatography   +   +
   c) Normal-phase chromatography   +   +
   d) Reversed-phase chromatography   +   +
   e) Ion-exchange chromatography   +   +

Advantages are indicated by the number of crosses (+).

Table II. Tentative ratio of conventional to microbore column chromatography in various types of laboratories.

N	Type of laboratory	Conventional chromatography (%)	Microbore column chromatography (%)
1.	Laboratories with a great extent of preparative work in chemistry, bioorganic chemistry, preparative chemistry	60	40
2.	Analytical biochemistry	20	80
3.	Clinical chemistry	10	90
4.	Agricultural chemistry	20	80
5.	Industrial chemistry without using very expensive eluents	80	20
6.	Industrial chemistry with the use of expensive eluents, analysis of trace amounts	20	80
7.	Environmental protection	10	90

LIST OF SYMBOLS

α — selectivity of separation;

β — void fraction of the column cross-section;

C — critical value of h/ν , corresponding to the begining of plateu region of $h/\nu = f(\nu)$ dependence;

c^{*} — the smallest detectible concentration of the solute;

d — diffusion distance in LC: $d = d_p$ for packed columns, $d = d_c$ for open tubular capillary columns;

d_p — particle diameter;

d_c — column diameter;

d_{opt} — optimum value of d;

D_m — diffusion coefficient of the solute in the mobile phase;

η — dynamic viscosity;

h — reduced plate height;

h_c — critical value of h, corresponding to $h/\nu = C$;

h_{min} — minimum value of the reduced plate height;

H — plate height or height equivalent to a theoretical plate;

k' — capacity factor of the solute;

k'_n — capacity factor for the last peak;

k_o — hydrodynamic permeability parameter of the column;

K_R — separation factor;

L — column length;

n — peak capacity - number of peaks which may be resolved on the column with efficiency N;

N — efficiency - number of theoretical plates;

N^{+} — minimum efficiency of open tubular capillary column, which is compatible with a detector with cell volume V_d;

ν - reduced velocity of the mobile phase;

ν_c - critical value of ν, corresponding to $h/\nu = C$;

ν_{opt} - the value of ν, corresponding to h_{min};

ΔP - pressure gradient along the column;

ΔP_c - critical value of ΔP, corresponding to $h/\nu = C$;

ΔP_{max} - maximum pressure in the column;

ΔP^* - fixed value of ΔP;

σ_v - standard deviation in volume units of the eluted peak;

$\sigma_{v,min}$ - minimum value of standard deviation;

t_R - time to achieve the given extent of separation;

$t_{R,C}$ - critical value of t_R, corresponds to $h/\nu = C$;

t_R^* - fixed value of t_R;

V_c - column volume;

V_d - detector cell volume;

V_d - maximum acceptable cell volume of the detector for $h/\nu = C$.

LITERATURE

1. Shimadzu High Performance Liquid Chromatograph, Model LC-5A, catalog No. 197-906-

2. P. Kucera, G. Manius, J. Chromatogr. 216, 9-21 (1981).

3. R. P. W. Scott, P. Kucera, M. Munroe, J. Chromatogr. 186, 475-487 (1979).

4. Jasco Model Familic 300 High Performance Semi-Micro Liquid Chromatograph, Japan Spectroscopic Co., Ltd., Catalog No. C-593-8205.

5. T. Takeuchi, D. Ishii, J. Chromatogr. 213, 25-33 (1981).

6. R. Tijssen, J. P. A. Bleumer, A. L. C. Smit, M. E. Van Kreveld, J. Chromatogr. 218, 137-165 (1981).

7. G. Guiochon, J. Chromatogr. 185, 3-26 (1979).

8. T. Tsuda, M. Novotny, Anal. Chem. 50, 271-275 (1978).

9. Zhidkostny Khromatograph (Liquid Chromatograph), "Milichrom" USSR Nauchpribor, (City-Orel).

10. Jasco Model Familic 100 N High-performance Micro Liquid Chromatograph, Japan Spectroscopic Co., Ltd., Catalog No. C 582-8107.

11. J. C. Giddings, "Dynamics of chromatography", M. Dekker, New York, 1965.

12. J. H. Knox, M. Salem, J. Chromatogr. 7, 614-622 (1969).

13. M. J. E. Golay in "Gas Chromatography" (D. H. Desty, ed.) Butterworths London, p. 36, 1958.

14. R. P. W. Scott, Trace Organic Analysis, Proc. 9-th Res. Symp., U. S. Gov. Printing Office, Washington, D. C., p. 637 (National Bureau of Standards special publication No. 519) 1979.

15. K. Slais, M. Krejci, J. Chromatogr. <u>235</u>, 21-29 (1982).

16. K. H. Schaffer, H. Levsen, J. Chromatogr. <u>206</u>, 245-252 (1981).

17. L. W. Hershberger, J. B. Callis, G. D. Christian, Anal. Chem. 51, 1444-1446 (1979).

18. I. Halász, R. Endele, J. Asshauer, J. Chromatogr. <u>112</u>, 37-60 (1975).

19. E. S. Yeung, M. J. Sepaniak, Anal. Chem. <u>52</u>, 1465A-1481A (1980).

20. E. M. Koroleva, V. G. Maltsev, B. G. Belenkii, J. Chromatogr. <u>242</u>, 145-152 (1982).

21. M. Goto, Y. Koyanagi, D. Ishii, J.Chromatogr. <u>208</u>, 261-262 (1981).

22. M. A. Berezhkovskii, B. G. Belenkii, R. G. Vinogradova, M. B. Ganitskii, N. E. Zhiltsova, V. G. Maltsev, O. P. Shilov in "Pribori dlya nauchnykh issledovanii i avtomatizatsii experimenta (Instrumentation for Sci. Res. and Automatization of experiments), ed. V. A. Pavlenko, "Nauka", p. 169, 51, 1980.

23. R. P. W. Scott, P. Kucera, J. Chromatogr. <u>169</u>, 51-72 (1979).

24. M. Krejci, K. Tesarik, J. Pajurek, J. Chromatogr. <u>191</u>, 17-23 (1980).

25. J. J. Kever, B. G. Belenkii, E. S. Gankina, L. Z. Vilenchik, U. I. Kurenbin, T. P. Zhmakina, J. Chromatogr. <u>207</u>, 145-147 (1981).

26. H. W. Osterhoudt, J. W. Williams, J. Phys. Chem. <u>69</u>, 1050 (1965).

LIST OF CONTRIBUTORS

ÁBRAHÁM, M. (591)
   Department of Biochemistry, József Attila University,
   H-6726 Szeged, Közép fasor 52, Hungary

ANDREEV, S.V. (311)
   Institute of Pure Biochemicals of the Main Board of Micro-
   biological Industry, Pudozhskaja 7, Leningrad, 197110, USSR

BALÁSPIRI, L. (257)
   Department of Medical Chemistry, University Medical School,
   Szeged, Hungary

BARBARO, A.M. (719)
   Instituto di Farmacologia, Università di Bologna, Bologna,
   Italy

BARCELÓ, D. (697)
   Department of Analytical Chemistry, Faculty of Chemistry,
   University of Barcelona, Diagonal 647, Barcelona-28, Spain

BARDÓCZ, S. (187)
   Department of Biochemistry, University Medical School,
   P.O. Box 6, H-4012 Debrecen, Hungary

BÁRDOS, L. (359)
   Department of Animal Physiology, University of Agricultural
   Sciences, H-2103 Gödöllő, Hungary

BARTHOLEYNS, J. (163)
   Centre de Recherche MERRELL International, 16, rue d'Ankara
   67084 Strasbourg, Cedex, France

BASKOVSKY, V.E.   (517)
    Institute of Macromolecular Compounds of the USSR Academy
    of Sciences, Leningrad, USSR

BELENKII, B.G.   (517, 841)
    Institute of Macromolecular Compounds of the USSR Academy
    of Sciences, Leningrad, USSR

BEREZKIN, V.G.   (35)
    Institute A.V. Topchiev of Petrochemical Synthesis, USSR
    Academy of Sciences, 29 Leninsky Pr., Moscow, USSR

BIAGI, G.L.   (719)
    Istituto di Farmacologia, Università di Bologna, Bologna,
    Italy

BIDLÓ-IGLÓY, M.   (413)
    Institute for Drug Research, P.O. B. 82, H-1325, Budapest,
    Hungary

BIHARI, K.   (219)
    Institute of Neurology, Semmelweis University of Medicine,
    H-1083 Budapest, Balassa utca 6, Hungary

BOJARSKI, J.   (403)
    Department of Organic Chemistry, Nicolaus Copernicus
    Academy of Medicine, 30-048 Kraków, Poland

BORDÁS, B.   (601, 751)
    Plant Protection Institute of the Hungarian Academy of
    Sciences, Budapest, Hungary

BOREA, P.A.   (719)
    Istituto di Farmacologia, Università di Ferrara, Ferrara,
    Italy

BOROSS, F.   (673)
    Central Food Research Institute, Budapest, Hungary

BOROSS, L.   (591)
    Department of Biochemistry, József Attila University,
    H-6727 Szeged, Közép fasor 52, Hungary

BUSZEWSKI, B.   (421)
    Department of Chemical Physics, Chemistry Institute, Maria
    Curie-Sklodowska University, Lublin, Poland

BUTOV, Yu. S. (761)
Hospital Therapy Department No.2, N.A. Semashko Moscow
Medical Stomatologic Institute, 20/1, Delegatskaya Ul.,
103473 Moscow, USSR

CANTELLI FORTI, G. (719)
Istituto di Farmacologia, Università di Bologna, Bologna,
Italy

COPIKOVÁ, J. (809)
Department of Saccharide Chemistry and Technology,
Institute of Chemical Technology, 166 28 Praha 6, 3,
Suchbátarova, Czechoslovakia

CSERHÁTI, T. (601, 751)
Plant Protection Institute of the Hungarian Academy of
Sciences, Budapest, Hungary

DAMASIEWICZ, B. (679)
Faculty of Pharmacy, Medical Academy, 80-416 Gdańsk, Poland

DAVILA-HUERTA, G. (227)
Department of Dermatology, Yale University School of
Medicine, 333 Cedar Street, New Haven, Connecticut 06510,
USA

DÉVAI, M. (601)
Department of Applied Chemistry, Technical University of
Budapest, Budapest, Hungary

DÉVÉNYI, T. (535)
Institute of Enzymology, Biological Research Center of the
Hungarian Academy of Sciences, Budapest, Hungary

DINYA, Z. (445)
Institute of Organic Chemistry, Kossuth Lajos University,
Debrecen, Hungary

EDMINSON, P.D. (237)
Pediatric Research Institute, National Hospital of Norway,
Oslo 1, Norway

EEK, L. (697)
Derivados Forestales, S.A., P^O Sant Joan 15, Barcelona-10,
Spain

EFIMOVA, I.I.   (517)
    Institute of Macromolecular Compounds of the USSR Academy
    of Sciences, Leningrad, USSR

EKIERT, E.   (403)
    Department of Organic Chemistry, Nicolaus Copernicus
    Academy of Medicine, 30-048 Kraków, Poland

EKSTEEN, R.   (51)
    Supelco Inc., Supelco Park, Bellefonte, PA 16823, USA

ELŐDI, P.   (187)
    Department of Biochemistry, University Medical School,
    P.O. Box 6, H-4012 Debrecen, Hungary

FEHÉR, ZS.   (19)
    Institute for General and Analytical Chemistry, Technical
    University of Budapest, Gellért tér 4, 1111 Hungary

FEKETE, J.   (195)
    Institute for General and Analytical Chemistry, Technical
    University, Budapest, Hungary

FODOR, G.   (445)
    Alkaloida Chemical Works, Tiszavasvári, Hungary

FODOR, I.   (445)
    Alkaloida Chemical Works, Tiszavasvári, Hungary

FODOR, M.   (445)
    Alkaloida Chemical Works, Tiszavasvári, Hungary

GAIKWAD, A.G.   (817)
    Department of Chemistry, Indian Institute of Technology,
    Bombay-400 076, India

GALCERÁN, M.T.   (697)
    Department of Analytical Chemistry, Faculty of Chemistry,
    University of Barcelona, Diagonal 647, Barcelona-28, Spain

GANKINA, E.S.   (517, 841)
    Institute of Macromolecular Compounds of the USSR Academy
    of Sciences of the USSR, Leningrad, USSR

GAZDAG, M.   (467)
    Chemical Works of Gedeon Richter, Ltd., H-1475 Budapest 10,
    P.O.B. 27, Hungary

GLÖCKNER, G.   (41)
    Technical University of Dresden, Department of Chemistry,
    8027 Dresden, Mommsenstrasse 13, GDR

GÖNDÖS, GY.   (113)
    Institute of Organic Chemistry, József Attila University,
    Dóm tér 8, H-6720 Szeged, Hungary

GRÓF, J.   (377)
    Joint Research Organization of the Hungarian Academy of
    Sciences and the Semmelweis University of Medicine,
    Budapest, Hungary

GRZYBOWSKI, J.   (679)
    Faculty of Pharmacy, Medical Academy, 80-416 Gdańsk, Poland

GUERRA, M.C.   (719)
    Istituto di Farmacologia, Università di Bologna, Bologna,
    Italy

GUOTH, J.   (367, 377)
    2nd Department of Obstetrics and Gynecology, Semmelweis
    University of Medicine, Budapest, Hungary

HÁRSING Jr., L.G.   (203, 219)
    Institute of Experimental Medicine, Hungarian Academy of
    Sciences, 1450 Budapest, Szigony utca 43, Hungary

HORVAI, G.   (19)
    Institute for General and Analytical Chemistry, Technical
    University of Budapest, Gellért tér 3, 1111 Hungary

HORVÁTH, CS.   (11)
    Department of Chemical Engineering, Yale University,
    New Haven, CT 06520, USA

HUDECZ, F.   (273)
    Research Group for Peptide Chemistry, Hungarian Academy of
    Sciences, Múzeum krt. 4/B, H-1088 Budapest, Hungary

IDEI, M.   (367, 377)
    Joint Research Organization of the Hungarian Academy of
    Sciences and the Semmelweis University of Medicine,
    Department of Clinical Biochemistry, Budapest, Hungary

ISHCHENKO, A.M.   (311)
    Institute of Pure Biochemicals of the Main Board of Micro-
    biological Industry, Pudozhskaja 7, Leningrad, 197110, USSR

JANÁKY, T.   (287)
    Endocrine Unit, First Department of Medicine, University
    Medical School Szeged, 6720 Szeged, Dóm tér 8, Hungary

JÁNOS, É.   (751)
    Research Institute for Plant Protection, Hungarian Academy
    of Sciences, Budapest, Herman Ottó u. 15, 1022 Hungary

KALÁSZ, H.   (195, 501, 535)
    Department of Pharmacology, Semmelweis University of
    Medicine, Budapest, Hungary

KALISZAN, R.   (679)
    Faculty of Pharmacy, Medical Academy, 80-416 Gdańsk, Poland

KARSAI, T.   (187)
    Department of Biochemistry, University Medical School,
    P.O. Box 6, H-4012 Debrecen, Hungary

KERECSEN, L.   (195)
    Department of Pharmacology, Semmelweis University of
    Medicine, Budapest, Hungary

KETTRUP, A.   (143)
    Universität-GH Paderborn, Applied Chemistry, D-4790
    Paderborn, FRG

KHOPKAR, S.M.   (817)
    Department of Chemistry, Indian Institute of Technology,
    Bombay-400 076, India

KICINSKI, H.G.   (143)
    Universität-GH Paderborn, Applied Chemistry, D-4790
    Paderborn, FRG

KITAEVA, L.P. (633)
   V.J. Vernaski Institute of Geochemistry and Analytical
   Chemistry USSR Academy of Sciences, Moscow, USSR

KNÖDGEN, B. (163)
   Centre de Recherche Merrell International, 16, rue d'Ankara,
   67084 Strasbourg, Cedex, France

KNOLL, J. (1, 195, 337)
   Department of Pharmacology, Semmelweis University of
   Medicine, Budapest, Hungary

KOLOMIETS, L.N. (35)
   Institute A.V. Topchiev of Petrochemical Synthesis, USSR
   Academy of Sciences, 29 Leninsky Pr., Moscow, USSR

KOL'TSOV, P.A. (761)
   Hospital Therapy Department No.2, N.A. Semashko Moscow
   Medical Stomatologic Institute, 20/1, Delegatskaya Ul.,
   103473 Moscow, USSR

KOROBITSIN, L.P. (311)
   Institute of Pure Biochemicals of the Main Board of Micro-
   biological Industry, Pudozhskaja 7, Leningrad, 197110, USSR

KOROLEV, A.A. (35)
   Institute A.V. Topchiev of Petrochemical Synthesis, USSR
   Academy of Sciences, 29 Leninsky Pr., Moscow, USSR

KOVÁCS, K. (287)
   Department of Medical Chemistry, University Medical School
   Szeged, 6720 Szeged, Dóm tér 8, Hungary

KURCZ, M. (435, 459)
   Research Laboratory of Clinical Biochemistry, CHINOIN,
   Budapest, Hungary

KURENBIN, O.I. (841)
   Institute of Macromolecular Compounds of the USSR Academy
   of Sciences, Leningrad, USSR

KUSZMANN, A. (601)
   Department of Inorganic Chemistry, Technical University of
   Budapest, Budapest, Hungary

KVASNICKA, F. (809)
Department of Saccharide Chemistry and Technology,
Institute of Chemical Technology, 166 28 Praha 6, 3,
Suchbátarova, Czechoslovakia

LAMPARCZYK, H. (679)
Faculty of Pharmacy, Medical Academy, 80-416 Gdańsk, Poland

LEHMANN, H. (833)
Institute of Biochemistry of Plants, Academy of Sciences of
the GDR, Halle (Saale), Weinberg, GDR

LITVINOVA, L.S. (517)
Institute of Macromolecular Compounds of the USSR Academy
of Sciences, Leningrad, USSR

LODKOWSKI, R. (421)
Department of Chemical Physics, Chemistry Institute, Maria
Curie-Sklodowska University, Lublin, Poland

MAASFELD, W. (143)
Universität-GH Paderborn, Applied Chemistry, D-4790
Paderborn, FRG

MAGYAR, K. (203, 391)
Department of Pharmacodynamics, Semmelweis University of
Medicine, 1445 Budapest, Nagyvárad tér 4, Hungary

MALINOWSKA, I. (615)
Institute of Chemistry, M. Curie-Sklodowska University,
20-031 Lublin, Poland

MALTSEV, V.G. (841)
Institute of Macromolecular Compounds of the USSR Academy
of Sciences, Leningrad, USSR

MANCAS, D.GH. (655, 659, 665)
Institute of Hygiene and Public Health, Str. V. Babes, 14,
Jassy, 6600 - Romania

MATKOVICS, B. (331)
Biological Isotope Laboratory, József Attila University,
Szeged, Hungary

MATUS, Z. (129)
   "Labor" Instrument Works, H-1445 Budapest, P.O. Box 280,
   Hungary

MAZSAROFF, I. (337)
   Department of Pharmacology, Semmelweis University of
   Medicine, Budapest, Hungary

MEISNER, I.S. (761)
   Biochemistry Department, Central Research Laboratory, N.I.
   Pirogov 2nd Moscow Medical Institute, 1, Ostrovitianova ul.,
   117437 Moscow, USSR

MENYHÁRT, J. (367, 377)
   Joint Research Organization of the Hungarian Academy of
   Sciences and the Semmelweis University of Medicine,
   Department of Clinical Biochemistry, Budapest, Hungary

MOKROSZ, J. (403)
   Department of Organic Chemistry, Nicolaus Copernicus
   Academy of Medicine, 30-048 Kraków, Poland

MUSIL, V. (809)
   Spolek por chemickou a hutni vyrobu, Usti n. Labem,
   Czechoslovakia

NAGY, J. (337)
   Department of Pharmacology, Semmelweis University of
   Medicine, Budapest, Hungary

NASAL, A. (679)
   Faculty of Pharmacy, Medical Academy, 80-416 Gdansk, Poland

OBRUBA, K. (107)
   Research Institute for Organic Syntheses, 532 18 Pardubice-
   Rybitví, Czechoslovakia

OCHÝNSKI, J. (421)
   Department of Toxicological Chemistry, Faculty of Pharmacy
   of Medical Academy, Lublin, Poland

OHMACHT, R. (119, 129)
   University Medical School, Institute of Chemistry, H-7643
   Pécs, P.O. Box 99, Hungary

ORR, J.C.   (113)
    Faculty of Medicine, Memorial University, St. John's,
    Newfoundland A1B 3V6, Canada

PAJOR, A.   (367, 377)
    2nd Department of Obstetrics and Gynecology, Semmelweis
    University of Medicine, Budapest, Hungary

PÁLOSI-SZÁNTHÓ, V.   (435)
    Research Laboratory of Clinical Biochemistry, CHINOIN,
    Budapest, Hungary

PANKOWSKI,   (679)
    Faculty of Pharmacy, Medical Academy, 80-416 Gdańsk, Poland

PASECHNIK, V.A.   (311)
    Institute of Pure Biochemicals of the Main Board of Micro-
    biological Industry, Pudozhskaja 7, Leningrad, 197110, USSR

PENKE, B.   (257, 287)
    Department of Medical Chemistry, University Medical School,
    Szeged, Hungary

PIETROGRANDE, M.C.   (719)
    Istituto di Chimica, Università di Ferrara, Ferrara, Italy

PIVOVAROV, A.M.   (311)
    Institute of Pure Biochemicals of the Main Board of Micro-
    biological Industry, Pudozhskaja 7., Leningrad, 197110,
    USSR

POLYÁK, B.   (591)
    Department of Biochemistry, József Attila University,
    H-6726 Szeged, Közép fasor 52, Hungary

POLYKOVSKAYA, O. YA.   (761)
    Hospital Therapy Department No. 2. N.A. Semashko Moscow
    Medical Stomatologic Institute, 20/1. Delegatskaya Ul.,
    103473 Moscow, USSR

PONIEWAZ, M.   (615)
    Institute of Chemistry, M. Curie-Sklodowska University,
    20-031 Lublin, Poland

PUNGOR, E. (19)
    Institute for General and Analytical Chemistry, Technical
    University of Budapest, Gellért tér 4, 1111 Hungary

RAFEL, J. (731)
    S.A. Cros, Research Center, Badalona (Barcelona), Spain

REICHELT, K.L. (237)
    Pediatric Research Institute, National Hospital of Norway,
    Oslo 1, Norway

RIVIER, J. (287)
    Peptide Biology Laboratory, The Salk Institute, La Jolla,
    California, USA

ROZYLO, J.K. (615)
    Institute of Chemistry, M. Curie-Sklodowska University,
    20-031 Lublin, Poland

RUBINSTEIN, R.N. (633)
    V.J. Vernaski Institute of Geochemistry and Analytical
    Chemistry, USSR Academy of Sciences, Moscow, USSR

SALVADORI, S. (719)
    Istituto di Chimica Farmaceutica, Università di Ferrara,
    Ferrara, Italy

SÁROSI, P. (367)
    Department of Obstetrics and Gynecology, New York
    University School of Medicine, New York, USA

SEILER, N. (163)
    Centre de Recherche Merrel International, 16, rue d'Ankara,
    67084 Strasbourg, Cedex, France

SIMON, G. (751)
    Institute of Pathophysiology, Semmelweis University of
    Medicine, Budapest, Hungary

STAHL, E. (497)
    Department of Pharmacognosy and Analytical Phytochemistry,
    University of Saarland, Saarbrücken, FRG

SUPRYNOWICZ, Z. (421)
    Department of Chemical Physics, Chemistry Institute, Maria
    Curie-Sklodowska University, Lublin, Poland

SZABÓ, L. (331)
    Biological Isotope Laboratory, József Attila University,
    Szeged, Hungary

SZEPESI, G. (467)
    Chemical Works of Gedeon Richter, Ltd., H-1475 Budapest 10,
    P.O.B. 27, Hungary

SZÓKÁN, GY. (257, 273)
    Institute of Organic Chemistry, Eötvös Loránd University,
    H-1088 Budapest, Hungary

TARCALI, J. (195)
    Department of Pharmacology, Semmelweis University of
    Medicine, Budapest, Hungary

TÁRCZY, M. (219)
    Institute, of Neurology, Semmelweis University of Medicine,
    H-1083 Budapest, Balassa utca 6, Hungary

TEKES, K. (203)
    Department of Pharmacodynamics, Semmelweis University of
    Medicine, 1445 Budapest, Nagyvárad tér 4, Hungary

THOMA, J.J. (51)
    South Bend Medical Foundation, 530 North Lafayette
    Boulevard, South Bend, IN 46601, USA

TIKHONOV, YU. V. (153, 761)
    Biochemistry Department, Central Research Laboratory,
    Moscow 117437, Ostrovitianova, 1, USSR

TOGUZOV, R.T. (153, 761)
    Biochemistry Department, Central Research Laboratory,
    Moscow 117437, Ostrovitianova, 1, USSR

TOMATIS, R. (719)
    Istituto di Chimica Farmaceutica, Università di Ferrara,
    Ferrara, Italy

TÖRÖK, A. (257)
    Department of Medical Chemistry, University Medical School,
    Szeged, Hungary

TÓTH, G. (287)
    Department of Medical Chemistry, University Medical School
    Szeged, 6720 Szeged, Dóm tér 8, Hungary

TÓTH, GY. (119)
    University Medical School, Institute of Chemistry, H-7643
    Pécs, P.O. Box 99, Hungary

TÓTH, K. (19)
    Institute for General and Analytical Chemistry, Technical
    University of Budapest, Gellért tér 4, 1111 Hungary

TÓTH, KATALIN (435)
    Research Laboratory of Clinical Biochemistry, CHINOIN,
    Budapest, Hungary

TÓTH-MÁRKUS, M. (673)
    Central Food Research Institute, Budapest, Hungary

UI, N. (299)
    Department of Physical Biochemistry, Institute of
    Endocrinology, Gunma University, Maebashi, Japan

URBÁN SZABÓ, K. (459)
    Laboratory of Clinical Biochemistry, CHINOIN, Budapest,
    Hungary

VALKÓ, K. (739)
    Institute of Enzymology, Biological Research Center,
    Hungarian Academy of Sciences, P.O. Box 7, Budapest,
    1502 Hungary

VÁRADY, L. (337)
    Department of Pharmacology, Semmelweis University of
    Medicine, Budapest, Hungary

VARGA, J.M. (227)
    Department of Dermatology, Yale University School of
    Medicine, 333 Cedar Street, New Haven, Connecticut 06510,
    USA

VIZI, E.S. (203, 219)
    Institute of Experimental Medicine, Hungarian Academy of
    Sciences, 1450 Budapest, Szigony utca 4, Hungary

VOLYNETS, M.P.   (633)
    V.J. Vernaski Institute of Geochemistry and Analytical
    Chemistry, USSR Academy of Sciences, Moscow, USSR

WULFSON, A.N.   (63)
    Shemyakin Institute of Bioorganic Chemistry, USSR Academy
    of Sciences, 117988 Moscow B-334, Vavilova 32, USSR

YAKIMOV, S.A.   (63)
    Shemyakin Institute of Bioorganic Chemistry, USSR Academy
    of Sciences, 117988 Moscow B-334, Vavilova 32, USSR